It's another Quality Book from CGP

This book is for 11-14 year olds.

Whatever subject you're doing it's the same
old story — there are lots of facts and you've just got
to learn them. KS3 Maths is no different.

Happily this CGP book gives you all that important
information as clearly and concisely as possible.

It's also got some daft bits in to try and make the whole
experience at least vaguely entertaining for you.

What CGP is all about

Our sole aim here at CGP is to produce the highest quality
books — carefully written, immaculately presented and
dangerously close to being funny.

Then we work our socks off to get them out to you
— at the cheapest possible prices.

Published by Coordination Group Publications Ltd.
Typesetting and layout by Coordination Group Publications

Key Stage Three Consultants:
Robert Gibson
Mary Gibson

Design Editor: Ruso Bradley

ISBN-10: 1 84146 030 3
ISBN-13: 978 1 84146 030 7

Groovy website: www.cgpbooks.co.uk
Jolly bits of clipart from CorelDRAW®
Printed by Elanders Hindson Ltd, Newcastle upon Tyne.

Contents

Multiples, Factors and Prime Factors

Multiples

> The <u>MULTIPLES</u> of a number are simply its <u>TIMES TABLE</u>:

E.g. the <u>multiples of 15</u> are   15   30   45   60   75   90   105   120 ...

Factors

> The <u>FACTORS</u> of a number are all the numbers that <u>DIVIDE INTO IT</u>.   There's a special way to find them:

Example 1:   *"Find ALL the factors of 20".*

Start off with 1 x the number itself, then try 2 x , then 3 x and so on, listing the pairs in rows like this. Try each one in turn and put a dash if it doesn't divide exactly. Eventually, when you get a number *repeated*, you *stop.*

Increasing by 1 each time

1 x 20
2 x 10
3 x -
4 x 5
5 x 4

> So the <u>FACTORS OF 20</u>
> are   <u>1,2,4,5,10,20</u>

This method guarantees you find them <u>ALL</u> — but *don't forget 1 and 20!*

Factors   Example 2:   *"Find the factors of 36".*

<u>Check each one in turn,</u>
to see if it divides or not.
Use your calculator if
you're not totally confident.

1 x 36
2 x 18
3 x 12
4 x 9
5 x -
6 x 6 → The 6 has *repeated* so *stop here.*

> So the <u>FACTORS of 36</u>
> are <u>1,2,3,4,6,9,12,18,36</u>

Finding Prime Factors — The Factor Tree

<u>Any number</u> can be broken down into <u>a string of</u> <u>PRIME NUMBERS</u> (see P.2) <u>all multiplied together</u> — this is called "Expressing it as a product of prime factors", and to be honest it's pretty tedious – but it's in the Exam, <u>and it's not difficult so long as you know what it is.</u>

The mildly entertaining "Factor Tree" method is best, where you start at the top and split your number off into factors as shown.   Each time you get a prime, you <u>ring it</u> and you finally end up with <u>all the prime factors</u>, which you can then arrange <u>in order.</u>

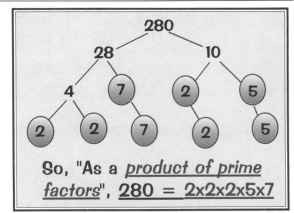

So, "As a *product of prime factors*", <u>280 = 2x2x2x5x7</u>

The Acid Test:
LEARN what <u>Multiples, Factors and Prime Factors</u> are, AND HOW TO FIND THEM. <u>Turn over and write it all down.</u>

Then try these <u>without looking at your notes</u>:
1) List the first 10 multiples of 8, and the first 10 multiples of 11.
2) List <u>all</u> the factors of 24 and <u>all</u> the factors of 60.
3) Express as a product of prime factors: a) 350   b) 480.

Prime Numbers

1) Basically, PRIME Numbers don't divide by anything

And that's the best way to think of them.

So Prime Numbers are all the numbers that DON'T come up in Times Tables:

| 2 | 3 | 5 | 7 | 11 | 13 | 17 | 19 | 23 | 29 | 31 | 37 | ... |

As you can see, they're an awkward-looking bunch (that's because they don't divide by anything!).   For example:

| The only numbers that multiply to give 5 are | 1×5 |
| The only numbers that multiply to give 23 are | 1×23 |

In fact the only way to get ANY PRIME NUMBER is $1 \times$ ITSELF

2) They All End in 1, 3, 7 or 9

1) 1 is NOT a prime number

2) The first four prime numbers are 2, 3, 5 and 7

3) 2 and 5 are the EXCEPTIONS because
all the rest end in 1, 3, 7 or 9

4) But NOT ALL numbers ending in 1, 3, 7 or 9
are primes, as shown here:

Only the circled numbers are primes

② ③ ⑤ ⑦
⑪ ⑬ ⑰ ⑲
21 ㉓ 27 ㉙
㉛ 33 ㊲ 39
㊶ ㊸ ㊼ 49

3) HOW TO FIND PRIME NUMBERS   — a very simple method

1) Since all primes (above 5) end in 1, 3, 7, or 9, then to find a prime number
between say, 50 and 60, the only possibilities are: 51, 53, 57 and 59

2) Now, to find which of them ACTUALLY ARE primes you only need to divide
each one by 3 and 7. If it doesn't divide exactly by either 3 or 7 then it's a prime.
(This simple rule using just 3 and 7 is true for checking primes up to 120)

So, to find the primes between 50 and 60, just try dividing 51, 53, 57 and 59 by 3 and 7:

$51 \div 3 = 17$, — 17 is a whole number,   so 51 is NOT a prime number
because it will divide by 3 ($3 \times 17 = 51$).

$53 \div 3 = 17.666$, $53 \div 7 = 7.571$ so 53 IS a prime number
because it ends in 1, 3, 7 or 9 and it doesn't divide by 3 or 7.

$57 \div 3 = 19$ — 19 is a whole number,   so 57 is NOT a prime number
because it will divide by 3 ($3 \times 19 = 57$).

$59 \div 3 = 19.666$ $59 \div 7 = 8.429$ so 59 IS a prime number.

The Acid Test: LEARN the main points in ALL 3 SECTIONS above.

Now cover the page and write down everything you've just learned.
1) Find all the prime numbers between 60 and 90 using the above method.
2) Make a number line from 1 to 50 and colour all the prime numbers (without looking them up).

Special Number Sequences

There are *FIVE special sequences* of numbers that you should *KNOW*:

1) EVEN NUMBERS …all Divide by 2

2 4 6 8 10 12 14 16 18 20 …

All *EVEN* numbers *END* in 0, 2, 4, 6 or 8
e.g. 64, 192, 1000, 518

2) ODD NUMBERS …DON'T divide by 2

1 3 5 7 9 11 13 15 17 19 21 …

All *ODD* numbers *END* in 1, 3, 5, 7 or 9
e.g. 121, 89, 403, 627

3) SQUARE NUMBERS:

(1x1) (2x2) (3x3) (4x4) (5x5) (6x6) (7x7) (8x8) (9x9) (10x10) (11x11) (12x12)

1 4 9 16 25 36 49 64 81 100 121 144…

3 5 7 9 11 13 15 17 19 21 23

Note that the DIFFERENCES between the square numbers are all the ODD numbers.

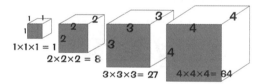

They're called *SQUARE NUMBERS* because they're like the areas of this pattern of squares:

1×1= 1 2×2= 4 3×3= 9 4×4= 16

4) CUBE NUMBERS:

(1x1x1) (2x2x2) (3x3x3) (4x4x4) (5x5x5) (6x6x6) (7x7x7) (8x8x8) (9x9x9) (10x10x10)…

1 8 27 64 125 216 343 512 729 1000…

1×1×1 = 1 2×2×2 = 8 3×3×3= 27 4×4×4= 64

They're called *CUBE NUMBERS* because they're like the volumes of this pattern of *cubes*.

Admit it, you never knew maths could be this exciting did you!

5) TRIANGLE NUMBERS:

To remember the triangle numbers you have to picture in your mind this *increasing pattern of triangles*, where each new row has *one more blob* than the previous row.

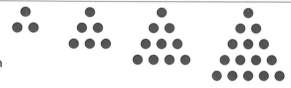

1 3 6 10 15 21 28 36 45 55 …

2 3 4 5 6 7 8 9 10 11 12

It's definitely worth learning this simple pattern of differences, as well as the formula for the nth term (see P.31) which is:

nth term = ½ n (n + 1)

The Acid Test:

LEARN the first 10 NUMBERS in all the five sequences: EVEN, ODD, SQUARE, CUBE and TRIANGLE NUMBERS.

1) Turn over and then write down the first 15 numbers in all five sequences.
2) Which number is even, a square number and a triangle number?
3) From this list of numbers: 27, 100, 1, 64, 49, 125, 16, 144, 25 write down:
   a) all the odd numbers; b) all the square numbers; c) all the cube numbers.

Equivalent Fractions

Equivalent fractions are fractions that are _equal in value_, even though they look different.

They Must Have the Same Multiplier Top and Bottom

Starting with any fraction you like, you can make up a list of equivalent fractions by simply _MULTIPLYING top and bottom_ by the _SAME NUMBER_ each time:

$$\frac{1}{2} = \frac{2}{4} = \frac{3}{6} = \frac{4}{8} = \frac{5}{10} \ldots\ldots = \frac{\text{any number}}{\text{TWICE that number}}$$

$$\frac{1}{5} = \frac{2}{10} = \frac{3}{15} = \frac{4}{20} = \frac{5}{25} \ldots\ldots = \frac{\text{any number}}{\text{FIVE TIMES that number}}$$

Cancelling Down

Going the other way, you will sometimes need to _simplify_ a fraction by _"cancelling down"_ — which only means _DIVIDING top and bottom_ by the _SAME NUMBER_:

$$\frac{3}{15} = \frac{1}{5} \qquad \div 3 \qquad \frac{22}{33} = \frac{2}{3} \qquad \div 11$$

Changing Fractions to Percentages and Back

If you can spot equivalent fractions then you can jump from fractions to percentages and back, nice and easily like this:

For example $\qquad \dfrac{1}{2} \qquad = \dfrac{5}{10} = \dfrac{50}{100} = 50\%$

and similarly $\qquad 20\% \qquad = \dfrac{20}{100} = \dfrac{2}{10} = \dfrac{1}{5}$

Equivalent fractions are good value for money. If you take a little bit of time and effort to sort them out, then you get all sorts of benefits in understanding percentages, solving equations, probability, an lots of other things. Equivalent fractions are real handy...

The Acid Test:

LEARN the three sections on this page. Make sure you know how to construct an equivalent fraction sequence.

By cancelling down, simplify the following as much as possible:

a) $\dfrac{12}{24}$; b) $\dfrac{8}{24}$; c) $\dfrac{6}{30}$; d) $\dfrac{6}{48}$; e) $\dfrac{5}{35}$; f) $\dfrac{7}{35}$; g) $\dfrac{25}{100}$; h) $\dfrac{39}{52}$.

Fractions, Decimals and Percentages

The one word that could describe all these three is <u>PROPORTION</u>. Fractions, decimals and percentages are simply <u>three different ways</u> of expressing a <u>proportion</u> of something — and it's pretty important you should see them as <u>closely related and completely interchangeable</u> with each other. This table shows the really common conversions which you should know straight off without having to work them out:

Fraction	Decimal	Percentage
1/2	0.5	50%
1/4	0.25	25%
3/4	0.75	75%
1/3	0.333333	33%
2/3	0.666667	67% rounded up
1/10	0.1	10%
2/10	0.2	20%
X/10	0.X	X0%
1/5	0.2	20%
2/5	0.4	40%

The more of those conversions you learn, the better — but for those that you <u>don't know</u>, you must <u>also learn</u> how to <u>convert</u> between the three types. These are the methods:

Fraction → Divide using the calculator e.g. ½ is 1÷2 = 0.5 → Decimal → × by 100 → Percentage e.g. 0.5 × 100 = 50%

Fraction ← The tricky one ← Decimal ← ÷ by 100 ← Percentage

<u>Converting decimals to fractions</u> is only possible for <u>exact decimals</u> that haven't been rounded off.

It's simple enough, but it's best illustrated by examples so look now at P.12 and work out what the simple rule is. You should then be able to fill in the rest of this table:

Fraction	Decimal	Percentage
3/4		
	0.2	
		70%
	0.55	
13/20		
	0.28	

The Acid Test:
LEARN the <u>whole of the top table</u> and the 4 conversion processes for FDP.

Now cover the page and write out the top FDP table from memory, and then the four conversion rules. Then fill in all the spaces in the 2nd table shown above.

Rounding Off

There are *two different ways* of specifying *where* a number should be *rounded off*.
They are: "Decimal Places" and "Significant Figures".   Doing "Decimal Places" is easier.

The question might say "*... to 4 DECIMAL PLACES*", or "*... to 3 SIGNIFICANT FIGURES*".
Don't worry, these are just different ways of *setting the position* of the *LAST DIGIT*.
Whichever way is used, the *basic method* is *always the same* and is *shown below*:

The Basic Method Has Three Steps

1) Identify the position of the LAST DIGIT.

2) Then look at the next digit to the RIGHT – called the DECIDER.

3) If the DECIDER is 5 or more, then ROUND-UP the LAST DIGIT.
If the DECIDER is 4 or less, then leave the LAST DIGIT as it is.

EXAMPLE: *"What is 8.35692 to 3 Decimal Places?"*

8.35692   = 8.357

LAST DIGIT to be written
(3rd decimal place because   *DECIDER*
we're rounding to 3 D P).

The *LAST DIGIT* rounds *UP*
because the *DECIDER* is *5*
or more.

Decimal Places   (D.P)

This is pretty easy:
  1) To round off to, say, 4 decimal places, the *LAST DIGIT* will be
                    the *4th one after the decimal point*.
  2) There must be *no more digits* after the LAST DIGIT (not even zeros).

DECIMAL PLACES EXAMPLES
Original number: 65.228371

Rounded to 5 decimal places (5 d p) 65.22837   (DECIDER was 1, so don't round up)
Rounded to 4 decimal places (4 d p) 65.2284   (DECIDER was 7, so do round up)
Rounded to 3 decimal places (3 d p) 65.228   (DECIDER was 3, so don't round up)
Rounded to 2 decimal places (2 d p) 65.23   (DECIDER was 8, so do round up)

The Acid Test:

LEARN the 3 Steps of the Basic Method and
the 2 Extra Points for Decimal Places.

Now turn over and write down what you've learned. Then try again till you know it.
1) Round 1.0672 to 2 decimal places.   2) Give 12.1566 to 2 decimal places.
3) Express 90.2532 to 3 DP.   4) Express 256.045 to 1 d.p.

Rounding Off

Significant Figures *(Sig. Fig.)*

The method for sig. fig. is *identical* to that for DP except that finding the *position* of the *LAST DIGIT* is more difficult — *it wouldn't be so bad, but for the ZEROS* ...

1) The 1st significant figure of any number is simply THE FIRST DIGIT WHICH ISN'T A ZERO.

2) The 2nd, 3rd, 4th, etc. significant figures follow on immediately after the 1st, REGARDLESS OF BEING ZEROS OR NOT ZEROS.

E.G. 0.003409   3.04070

SIG FIGS: 1st 2nd 3rd 4th   1st 2nd 3rd 4th
(If we're rounding to say, 3 sig. fig. then the LAST DIGIT is simply the 3rd sig. fig.)

3) After *Rounding Off* the LAST DIGIT, end ZEROS must be filled in up to, BUT NOT BEYOND, the decimal point.

No *extra zeros* must ever be put in *after* the decimal point.

Examples	*to 4 SF*	*to 3 SF*	*to 2 SF*	*to 1 SF*
1) 34.8751	34.88	34.9	35	30
2) 16.0057	16.01	16.0	16	20
3) 0.0023904	0.002390	0.00239	0.0024	0.002
4) 10795.2	10800	10800	11000	10000

POSSIBLE ERROR OF HALF A UNIT WHEN ROUNDING

Whenever a measurement is *rounded off* to a *given UNIT* the actual measurement can be anything up to **HALF A UNIT** bigger or smaller.

Examples:

1) A wall is given as being *"6m high to the nearest METRE"* — its actual height could be anything in-between *5.5m to 6.5m* — i.e. **HALF A METRE** either side of 6m.

2) If it was given as *"6.4m, to the nearest 0.2m"*, then it could be anything from *6.3m to 6.5m* — i.e. *0.1m either side* of 6.4m.

3) *"A school has 2800 pupils to 2 Sig Fig"* (i.e. to the nearest 100) — the actual figure could be anything *from 2750 to 2849*.   — (Why isn't it 2850?)

The Acid Test: LEARN the whole of this page, then turn over and write down everything you've learned. It's all good clean fun.

1) **Round these to 2 D.P:**   a) 7.309   b) 0.057   c) 1.083   d) 4.597
2) **Round these to 3 S.F,** and for each one say which of the 3 rules about ZEROS applies: a) 0.03582   b) 63615   c) 345.86   d) 0.70985
3) **A room is described as 18 feet long to the nearest foot. What is the longest and** shortest it could be, in feet and inches? (e.g. 13 feet 3 inches).

SECTION ONE — NUMBERS MOSTLY

Accuracy and Estimating

Appropriate Accuracy

In your Exam you may well get a question asking for _"an appropriate degree of accuracy"_ for a certain measurement.

So how do you decide what is _appropriate accuracy_? The key to this is _the number of significant figures_ (See P.7) that you give it to, and these are the simple rules:

1) For fairly <u>casual measurements, 2 SIGNIFICANT FIGURES</u> is most appropriate.

> <u>EXAMPLES</u>:
> <u>COOKING</u> — 250g (2 sig fig) of sugar,
> (_not_ 253g (3 S F), or 300g (1 S F))
> <u>DISTANCE OF A JOURNEY</u> — 320 miles or 15 miles or 2400 miles (All 2 S F)
> <u>AREA OF A GARDEN OR FLOOR</u> — 480m^2 or 18m^2

2) For <u>MORE IMPORTANT OR TECHNICAL THINGS, 3 SIGNIFICANT FIGURES</u> is essential.

> <u>EXAMPLES</u>:
> A <u>LENGTH</u> that will be <u>CUT TO FIT</u>, e.g. You'd measure a shelf as <u>**75.6cm**</u> long
> (_not_ <u>**76cm or 75.63cm**</u>)
> A <u>TECHNICAL FIGURE</u>, e.g. <u>**35.1**</u> miles per gallon,
> (_rather than_ 35 mpg)
> Any <u>ACCURATE</u> measurement with a ruler:
> e.g. <u>**44.5cm**</u>, (_not_ 40cm or 44.54cm)

3) Only for <u>REALLY SCIENTIFIC WORK</u> would you have <u>more than 3 SIG FIG</u>.

For example, only someone _really keen_ would want to know the length of a piece of pipe _to the nearest tenth of a mm_ — like 34.46cm, for example. (_Get a life!_)

Estimating Calculations

As long as you realise what's expected, this is _VERY EASY_. People get confused because they _over-complicate it_. To _estimate_ something this is all you do:

> 1) ROUND EVERYTHING OFF to nice easy CONVENIENT NUMBERS
> 2) Then WORK OUT THE ANSWER using those nice easy numbers
> — and that's it!

You don't worry about the answer being "wrong", because we're only trying to get a rough idea of the size of the proper answer, e.g. is it about 50 or about 500? Don't forget though, in the Exam you'll need to _show all the steps you've done_, to prove you didn't just use a calculator. Look at the example up there ↗

Accuracy and Estimating

An Example of Estimating the answer to a Calculation:

Q: ESTIMATE the value of $\dfrac{48.6 \times 5.2}{117.4 + 375.9}$ showing all your working.

ANSWER:

$$\frac{48.6 \times 5.2}{117.4 + 375.9} \approx \frac{50 \times 5}{120 + 380} \approx \frac{250}{500} \approx \frac{1}{2}$$ ("≈" means *roughly equal to*)

Estimating Areas and Volumes

This isn't bad either — so long as you *LEARN* the *TWO STEPS* of the method:

1) Draw or imagine a RECTANGLE OR CUBOID of similar size to the object in question.

2) Round off all lengths to the NEAREST WHOLE, and work it out — easy.

EXAMPLES:

a) Estimate the area of Australia:

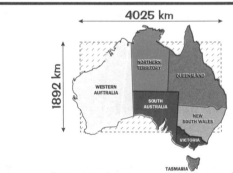

Area of Australia is *approximately equal to* area of dashed rectangle:
i.e. 1892km × 4025km = 7 615 300km²
(or, without a calculator:
2000 × 4000 = 8 000 000km²)

b) Estimate the volume of the vase:

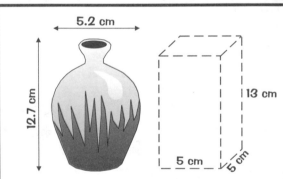

Volume of vase is *approximately equal to* volume of dashed cuboid
= 5 x 5 x 13
= 325cm³

The Acid Test:

LEARN the 3 RULES about Appropriate Accuracy and the 4 RULES for Estimating.

Then cover the page and write them all down from memory.

THEN TRY THESE:

1) Estimate: a) the area of the UK in square miles,
b) the volume of a football in inches³.

2) Decide which category of accuracy these should belong in and round them off accordingly:
a) A wedding cake weighs 2.561kg
b) A man is 182.61cm
c) A village has 1048 inhabitants
d) A train travels at 117.55mph.

SECTION ONE — NUMBERS MOSTLY

Conversion Factors

Conversion Factors are a very powerful tool for dealing with a wide variety of questions and the method is very easy.

Method

1) Find the <u>Conversion Factor</u> (always easy)

2) <u>Multiply by it AND divide by it</u> (to get 2 answers)

3) Choose which is the <u>common sense answer</u>

Three Important Examples

1) *"Convert 4.45 hours into minutes."*   (This is NOT 4hrs 45mins)

 1) Conversion factor = <u>60</u>   (simply because 1 hour = <u>60</u> mins)
2) 4.45 hrs × 60 = 267 mins (makes sense)
   4.45 hrs ÷ 60 = 0.0742 mins (ridiculous answer!)
3) So plainly the answer is that 4.45hrs = <u>267 mins</u>   (=4hrs 27mins)

2) *"If £1 = 2.75 Deutschmarks, how much is 15.43 Deutschmarks in £ and p?"*

1) Obviously, Conversion Factor = <u>2.75</u>   (The "exchange rate")
2) 15.43 × 2.75 = £42.43
   15.43 ÷ 2.75 = £5.61
3) Not quite so obvious this time, but if roughly 3 Deutschmarks = £1,
   then 15 Deutschmarks can't be much — certainly not £42,
   so the answer must be <u>£5.61p</u>

3) *"A map has a scale of 1:50,000. How big in real life is a distance of 6cm on the map?"*

1) Conversion Factor = 50 000
2) 6cm × 50 000 = 300 000cm (looks OK)
   6cm ÷ 50 000 = 0.00012cm (not good)
3) So 300 000cm is the answer.
   How do we convert to metres?

To Convert 60,000cm to m:
1) C.F. = 100   (cm ⟺ m)
2) 300 000 × 100 = 30,000,000m (hmm)
   300 000 ÷ 100 = <u>3000m</u> (more like it)
3) So answer = <u>3000m</u> (or 3km).

The Acid Test:

LEARN the <u>3 steps</u> of the <u>Conversion Factor method</u>.
Then turn over and <u>write them down</u>.

1) Convert 5.4 km into metres.
2) Which is more, £28 or 13 Deutschmarks? (Use 2.75)
3) A map is drawn to a scale of 4cm = 5km. A road is 40 km long. How many cm
   will this be on the map? (Hint, C.F. = 5÷4, i.e. 1cm = 1.25 km)

Metric and Imperial Units

This topic is *Easy Marks!* — make sure you get them.

Metric Units

1) <u>Length</u> mm, cm, m, km
2) <u>Area</u> mm², cm², m², km²,
3) <u>Volume</u> mm³, cm³, m³,
      litres, ml
4) <u>Weight</u> g, kg, tonnes
5) <u>Speed</u> km/h, m/s

MEMORISE THESE KEY FACTS:

1cm = 10mm	1 tonne = 1000kg
1m = 100cm	1 litre = 1000ml
1km = 1000m	1 litre = 1000cm³
1kg = 1000g	1 cm³ = 1 ml

Imperial Units

1) <u>Length</u> Inches, feet, yards, miles
2) <u>Area</u> Square inches, square feet, square yards, square miles
3) <u>Volume</u> Cubic inches, cubic feet, gallons, pints
4) <u>Weight</u> Ounces, pounds, stones, tons
5) <u>Speed</u> mph

LEARN THESE TOO!

1 Foot = 12 Inches
1 Yard = 3 Feet
1 Gallon = 8 Pints
1 Stone = 14 Pounds (lbs)
1 Pound = 16 Ounces (Oz)

Metric-Imperial Conversions

<u>YOU NEED TO LEARN THESE</u> — they DON'T promise to give you these in the Exam and if they're feeling mean (as they often are), they won't.

APPROXIMATE CONVERSIONS

1 kg = 2¼ lbs	1 gallon = 4.5 litres
1m = 1 yard (+ 10%)	1 foot = 30cm
1 litre = 1¾ pints	1 metric <u>tonne</u> = 1 imperial <u>ton</u>
1 inch = 2.5 cm	1 mile = 1.6km or 5 miles = 8 km

Using Metric-Imperial Conversion Factors   (See P.10)

1) *Convert 36mm into cm.*
   ANS: CF = 10, so × or ÷ by 10, which gives 360cm or <u>3.6cm</u>. (Sensible)
2) *Convert 45 inches into cm.*
   ANS: CF = 2.5, so × or ÷ by 2.5, which gives 18cm or <u>112.5cm</u>.
3) *Convert 2.77 litres into pints*
   ANS: CF = 1¾, so × or ÷ by 1.75, which gives 1.58 or <u>4.85 pints</u>.

The Acid Test:
LEARN THE <u>21 Conversion factors</u> in the boxes above then cover up the page and <u>write them all down</u>.

1) A car travels at 55 mph. What is its speed in km/h?
2) Roughly how many yards is 200m? 3) What is 74 inches in cm?
4) Petrol costs £3.14 per gallon. What should it cost per litre?
5) How many litres is 2½ gallons?

Fractions Without the Calculator!

Terrifyingly, they may force you (under Exam conditions!) to demonstrate your prowess at doing fractions _by hand_ ... better learn this little lot first then, eh!

1) Converting Fractions to Decimals — Just DIVIDE

Just remember that " / " means " ÷ ", <u>so ¼ means 1÷4 = 0.25</u>

A/B — What does it mean?

This is confusing unless you know it's just <u>a way of writing a FRACTION on a single line</u>, and you should know that a fraction actually means <u>ONE THING DIVIDED BY ANOTHER</u>.

So A/B, $\frac{A}{B}$, $^A/_B$ and A ÷ B _all mean the same thing._

For example, 5/11 , $\frac{5}{11}$, $^5/_{11}$ and 5 ÷ 11 (=0.4545....) are all the same thing.

2) Converting Decimals to Fractions
— it's a simple rule, so work it out yourself!:

0.2 $= ^2/_{10}$,   0.9 $= ^9/_{10}$,   0.6 $= ^6/_{10}$,   etc.

0.15 $= ^{15}/_{100}$,   0.32 $= ^{32}/_{100}$,   0.06 $= ^6/_{100}$,   0.68 $= ^{68}/_{100}$, etc.

0.234 $= ^{234}/_{1000}$,   0.085 $= ^{85}/_{1000}$,   0.505 $= ^{505}/_{1000}$,   etc.

These can then be _cancelled down_.

Doing it By Hand:

1) Multiplying — easy
Multiply top and bottom separately: $\frac{2}{3} \times \frac{5}{7} = \frac{2 \times 5}{3 \times 7} = \frac{10}{21}$

2) Dividing — quite easy
Turn the <u>2nd fraction UPSIDE DOWN</u> and then <u>multiply</u>: $\frac{1}{2} \div \frac{6}{7} = \frac{1}{2} \times \frac{7}{6} = \frac{7}{12}$

3) Adding , subtracting — fraught
Add or subtract _TOP LINES ONLY_ but _only if the bottom numbers are the same._
(If they're not the same it gets very tricky.)

$\frac{3}{5} + \frac{1}{5} = \frac{4}{5}$

$\frac{7}{8} - \frac{3}{8} = \frac{4}{8}$

4) Cancelling down — easy
Divide top and bottom by the same number, till they won't go any further:

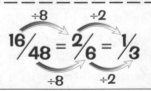

$^{16}/_{48} = ^2/_6 = ^1/_3$

Fractions With the Calculator

If at all possible, use your calculator to do all fractions in the Exam.

The Fraction Button:

Use this as much as possible in the Exam. It's very easy, so make sure you know how to use it — you'll lose a lot of marks if you don't:

1) TO ENTER A NORMAL FRACTION like $\frac{1}{5}$

Just press: **1** **$a^b/_c$** **5**

Walter always had trouble with calculators...

2) TO ENTER A MIXED FRACTION like $1\frac{3}{4}$

Just press: **1** **$a^b/_c$** **3** **$a^b/_c$** **4**

3) TO DO A REGULAR CALCULATION such as $\frac{2}{5} \times \frac{1}{4}$

Just press: **2** **$a^b/_c$** **5** **×** **1** **$a^b/_c$** **4** **=**

4) TO REDUCE A FRACTION TO ITS LOWEST TERMS

Just enter it and then press: **=**

e.g. $\frac{12}{24}$ · **12** **$a^b/_c$** **24** **=**   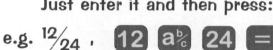 $= \frac{1}{2}$.

5) TO CONVERT BETWEEN MIXED AND TOP HEAVY FRACTIONS

Just press **SHIFT** **$a^b/_c$**   e.g. to give $3\frac{5}{7}$ as a top heavy fraction:

Press: **3** **$a^b/_c$** **5** **$a^b/_c$** **7** **=** , **SHIFT** **$a^b/_c$** which gives an answer of $\frac{26}{7}$.

The Acid Test:

LEARN the 2 Rules for converting Fractions ↔ Decimals, the 4 Manual Methods and the 5 features of the Fraction Button.

Then cover up these two pages and write down what you've learned.

1) Do these **BY HAND**:

   a) $\frac{1}{4} \times \frac{5}{7}$ b) $\frac{2}{3} \div \frac{5}{9}$ c) $\frac{5}{12} + \frac{3}{12}$ d) Express $\frac{33}{55}$ in its lowest terms.

2) Do these **WITH YOUR CALCULATOR**:

a) Convert $\frac{3}{5}$ into a decimal   b) Convert 0.88 into a fraction, and cancel it down.

c) $\frac{1}{2} \times \frac{3}{5}$ d) $\frac{5}{6} \div \frac{3}{4}$ e) $\frac{7}{8} - \frac{4}{8}$ f) Find x: $3\frac{1}{4} = \frac{x}{4}$ g) Find y: $\frac{36}{81} = \frac{4}{y}$.

Percentages

Contrary to popular belief there are *three distinct types* of percentage question. Obviously then, it's going to be pretty *essential* that you can:

> 1) Distinguish between the three types
> 2) Remember the METHOD for each of them

Type 1 — THESE ARE IDENTIFIED BY THE "%" SYMBOL IN THE QUESTION

This is the easiest type — they're always of the form:

> *FIND "something" % OF "something-else"*

Method

For example: Find 35% *of* £200

> 1) *WRITE:* 35% OF £200
>
> 2) *TRANSLATE:* $\frac{35}{100}$ × 200 = £70
>
> 3) *CHECK* THAT IT'S A *SENSIBLE ANSWER*.

Remember:

1) "OF" means "X"

2) "PER CENT" means "OUT OF 100", so 35% *means* "35 out of 100", i.e. $\frac{35}{100}$.

Type 2 — THESE ARE IDENTIFIED BY THE WORD *"PERCENTAGE"* IN THE QUESTION

These are always of the form:

> *EXPRESS "one thing" AS A PERCENTAGE OF "another"*

For example: Express £12 *as a percentage* of £80.

Method — FDP

F D P : Fraction – Decimal – Percentage

(See P.5) $\frac{12}{80}$ $\xrightarrow{12 \div 80}$ 0.15 $\xrightarrow{\times 100}$ 15%

Make a <u>fraction</u> using the two numbers — always with the <u>smallest on top</u>.

<u>Divide</u> them to get a <u>decimal</u>.

Then <u>multiply by 100</u> to get a <u>percentage</u>.

Percentages

Type 3 — THESE ARE IDENTIFIED BY THEM *NOT* GIVING YOU THE *"ORIGINAL VALUE"*

These are the type most people get wrong — but only because they don't recognise them as a type 3 and don't apply this simple method:

Example:

A house increases in value by 25% to £90,000. Find its value *before* the rise.

Method

	£90,000	=	125%
÷ 125	£720	=	1%
× 100	£72,000	=	100%

So the original price was £72,000

An *increase* of 25% means that £90,000 represents *125% of the original* value.

If it was a *DROP* of 25%, then we would put "*£90,000 = 75%*" instead, and then divide by *75* on the LHS, instead of 125.

Always set them out exactly like this example. The trickiest bit is deciding the top % figure on the RHS — the 2nd and 3rd rows are *always* 1% and 100%.

Percentage Change (*An important example of type 2*)

It's quite common to give a *change in value* as a *percentage.*
This is the formula for doing so — *LEARN IT, AND USE IT:*

$$\text{PERCENTAGE "CHANGE"} = \frac{\text{"CHANGE"}}{\text{ORIGINAL}} \times 100$$

By "change", we could mean all sorts of things such as: "Profit", "loss", "appreciation", "depreciation", "increase", "decrease", "error", "discount", etc. For example, *percentage "profit"* = $\frac{\text{"profit"}}{\text{original}}$ × 100.

Don't forget the monster importance of using the *ORIGINAL VALUE* in this formula.

The Acid Test: LEARN The 3 Types, how you identify them, and the Method for each. Also LEARN the Formula for Percentage Change.

Now *turn over and write down all the details* you've just learned.

Identify the following questions as Type 1, 2, or 3, and apply the method for each. *Practise until you can do them without the notes:*

1) Find the amount payable on a solicitor's bill given as: "£220 + 17.5% VAT".
2) A coat is reduced to £88 in a "20% off" sale. What was its pre-sale price?
3) A man's pay increases from £240 per week to £270. Find his weekly increase in £ and state it as a percentage.

Calculator Buttons 1

The next few pages are full of lovely calculator tricks to save you a lot of button-bashing. There's basically two types of calculator — the old-style and the new fancy two-line displayers.

The Old-Style Calculators:

These ones only display numbers. They do the calculation each time you press an operation key.

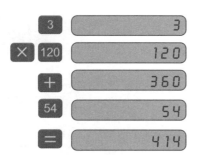

[C] *SEMI-CANCEL* and [AC] *ALL CANCEL*

The [C] button only cancels the NUMBER YOU ARE ENTERING. [AC] clears the whole calculation.

If you use [C] instead of [AC] for when you hit the wrong key, you'll HALVE the time you spend correcting mistakes!

2-line Display Calculators:

These fancy ones are dead common now. They're pretty easy to use because you just type most calculations exactly as they're written. Like this:

[DEL] *The Delete button*

Pressing the [DEL] button deletes what you've typed, one key at a time (just like on a computer), so it's much quicker than pressing [AC] and re-typing the whole lot. Use [DEL] or you'll be in BIG TROUBLE!

Cursor buttons [◄] [►]

These cursor buttons [◄] and [►] are pretty useful for editing what you've typed in. (You'll probably find you overwrite what was there before, but you can change this with the INS key to insert, rather than overwrite.)

1) Entering Negative Numbers

On some calculators, there's a [+/-] button. To enter a minus number, you need to press this after you've entered the number. A lot of calculators just have a minus button [(-)] which you press before entering the number.

So to work out – 5 × – 6 you'd either press...

or...

Why can't they all just be the same...
(The examples in this book will use the [(-)] button, but if yours is different make sure you know how to use it!)

Calculator Buttons 2

2) Square, Square Root and Cube Root

The SQUARE, SQUARE ROOT, and CUBE ROOT buttons are $\boxed{X^2}$ $\boxed{\sqrt{}}$ and $\boxed{\sqrt[3]{}}$.

1) The $\boxed{X^2}$ button <u>squares the number you typed</u>, i.e. <u>IT MULTIPLIES IT BY ITSELF</u>.
   It's ideal for finding the area of a circle, using the well-known (hah!) formula:
   $$A = \pi\, r^2 \qquad \text{e.g. if } r = 5 \text{ then press } \boxed{3.14}\ \boxed{\times}\ \boxed{5}\ \boxed{X^2}\ \boxed{=} \text{ which gives you 78.5.}$$
   (To get a more accurate answer, use the π button which is usually the second function of the EXP key)

2) $\boxed{\sqrt{}}$ is the <u>REVERSE PROCESS</u> of $\boxed{X^2}$ — it calculates the <u>SQUARE ROOT</u> of the
   number you enter. Pressing $\boxed{\sqrt{}}$ $\boxed{25}$ $\boxed{=}$ gives $\boxed{\qquad\qquad 5}$
   then $\boxed{X^2}$ $\boxed{=}$ takes you back to $\boxed{\qquad\quad 25}$.

3) $\boxed{\sqrt[3]{}}$ gives the <u>CUBE ROOT</u> (See P.20) which is the reverse of <u>CUBING</u> a number.
   E.g. $\boxed{\sqrt[3]{}}$ $\boxed{27}$ $\boxed{=}$ gives $\boxed{\qquad\qquad 3}$
   then pressing $\boxed{X^3}$ $\boxed{=}$ takes you back to $\boxed{\qquad\quad 27}$.

3) Older Calculators do stuff Backwards

On some calculators, especially older ones, you need to enter a lot of
calculations _backwards_. E.g. if you're working out the square root of a number, you'd
enter the number first and <u>then</u> press the square root.

$\boxed{25}$ $\boxed{\sqrt{}}$ $\qquad\qquad$ $\boxed{\qquad\qquad 5}$

Or if you were typing a trigonometry function like sin 45°, you'd do:

$\boxed{45}$ $\boxed{SIN}$ $\qquad$ $\boxed{0.70716718}$

You don't need to press equals when you do functions on one of these
calculators. It'll work it out automatically. Which is jolly nice of it.

4) The MEMORY BUTTONS ($\boxed{STO}$ Store, $\boxed{RCL}$ Recall)

(On some calculators the memory buttons are called $\boxed{Min}$ (memory in) and $\boxed{MR}$ (memory recall)).
Contrary to popular belief, the memory is not intended for storing your favourite
phone number, but in fact is a mighty useful feature for keeping a number you've just
calculated, so you can use it again shortly afterwards.

For something like $\dfrac{16}{15 + 12SIN40}$, you could just work out the _bottom line_ first
and _stick it in the memory_:

Press $\boxed{15}$ $\boxed{+}$ $\boxed{12}$ $\boxed{SIN}$ $\boxed{40}$ $\boxed{=}$ and then $\boxed{STO}$ (Or $\boxed{STO}$ $\boxed{M}$ or $\boxed{STO}$ $\boxed{1}$ or $\boxed{Min}$)
to keep the result of the bottom line in the memory.
Then you simply press $\boxed{16}$ $\boxed{\div}$ $\boxed{RCL}$ $\boxed{=}$, and the answer is 0.7044.
(Instead of $\boxed{RCL}$, you might need to type $\boxed{RCL}$ $\boxed{M}$ or $\boxed{RCL}$ $\boxed{1}$ or $\boxed{MR}$ on yours.)

Once you've practised with the memory buttons a bit, you'll soon find them very
useful. They can speed things up no end.

Calculator Buttons 3

5) *Bodmas and the Brackets Buttons*

The BRACKETS BUTTONS are (and).

One of the biggest problems people have with their calculators is not realising that the calculator always works things out IN A CERTAIN ORDER, which is summarised by the word BODMAS, which stands for:

Brackets, Other, Division, Multiplication, Addition, Subtraction

This becomes really important when you want to work out even a simple thing like

$\frac{23+45}{64\times3}$ — it's no good just pressing [23][+][45][÷][64][×][3][=] — it will be

completely wrong. The calculator will think you mean $23+\frac{45}{64}\times3$ because the calculator will do the *division and multiplication* BEFORE it does the *addition*.

The secret is to OVERRIDE the automatic BODMAS order of operations using the BRACKETS BUTTONS. Brackets are the ultimate priority in BODMAS, which means anything in brackets is worked out before anything else happens to it.
So all you have to do is

| 1) Write a couple of pairs of brackets into the expression: | $\frac{(23+45)}{(64\times3)}$ |
| 2) Then just type it as it's written: | |

[(][23][+][45][)][÷][(][64][×][3][)][=]

You might think it's difficult to know where to put the brackets in.
It's not that difficult, you just put them in pairs around each group of numbers.
It's OK to have brackets within other brackets too, *e.g. (4 + (5÷2))*
As a rule, you can't cause trouble by putting too many brackets in,
SO LONG AS THEY ALWAYS GO IN PAIRS.

6) *The Fraction Button:* [a b/c]

— It's absolutely essential that you learn how to use this button for doing fractions.
Full details are given on P.13.

7) *The Powers Button:* [x^y] On some calculators, the powers button is [^]

It's used for working out powers of numbers quickly. For example to find 7^5, instead of pressing 7×7×7×7×7 you should just press [7][x^y][5][=]

Calculator Buttons 4

8) The Standard Form Button

The STANDARD FORM BUTTON is [EXP] or [EE].

All you ever use this for is entering numbers written in *standard form* into the calculator.

It would be a lot more helpful if the calculator manufacturers labelled it as [x10^n] because <u>that's what you should call it as you press it</u>: *"Times ten to the power..."*

For example to enter 6×10^3 **YOU MUST ONLY PRESS:** [6] [EXP] [3]
and **NOT**, as a lot of people do: [6] [X] [10] [EXP] [3].
The calculator would then display this number as: [6^3]
(REMEMBER — it doesn't mean 6^3 but 6×10^3)

Pressing [X] [10] as well as [EXP] is <u>HORRIBLY WRONG</u>, because the [EXP] <u>ALREADY CONTAINS</u> the "× 10" in it. That's why you should always say to yourself "<u>TIMES TEN TO THE POWER</u>..." every time you press the [EXP] button, to prevent this very common mistake.

9) Modes

This is tricky and you wouldn't really need to know about it except that you'll sometimes accidentally get into the wrong mode, and it can make life pretty difficult if you don't know how to get back to normality.

There are <u>3 SEPARATE MODES</u> that your calculator has to make a choice about:

CALCULATION MODES
You want COMP mode. This is the mode for doing normal calculations.
On CASIOs, this is on the first menu you get from pressing [MODE]

ANGLES MODES
You want degrees mode (there'll be a small DEG or D on the display when you're in this mode).
On CASIOs, you'd press [MODE] twice to get the right menu.

DISPLAY MODES
You want NORM mode most of the time. The other display modes are for showing a certain number of decimals places (FIX) or number of significant figures (SCI).
(have a play with these — they're great fun... err, I mean they might be useful... or something.)

The Acid Test:

LEARN YOUR CALCULATOR BUTTONS. <u>PRACTISE</u> until you can do <u>all of these</u> without having to refer back:

1) What do the [x²] and [√] buttons do?
2) What must you press to find 17^2?
3) How do you enter -5 × -8?
4) Explain what [STO] and [RCL] do and give an example of using them.
5) What is the [a^b_c] button used for?
6) How do you enter 6^8?
7) How do you enter 6×10^8?
8) Which should be showing at the top of your display: DEG, RAD or GRAD?

Ratios

The whole grisly subject of <u>RATIOS</u> gets a whole lot easier when you do this:

Treat RATIOS like FRACTIONS

So for the <u>RATIO</u> 3:4, you'd treat it as the <u>FRACTION</u> 3/4, which is 0.75 as a <u>DECIMAL</u>.

What the fraction form of the ratio actually means

Suppose in a class there's <u>girls and boys</u> in the ratio 3 : 4.
This means there's 3/4 as many girls as boys.
So if there were 20 boys, there would be 3/4 × 20 = 15 girls.
You've got to be careful — it <u>doesn't mean</u> 3/4 of the <u>people</u> in the class are girls.

Reducing Ratios to their simplest form

You reduce ratios just like you'd reduce fractions to their simplest form.

For the ratio 15 : 18, both numbers have a <u>factor</u> of 3, so <u>divide them by 3</u> —
That gives 5 : 6. We can't reduce this any further. So the simplest form of 15 : 18 is <u>5 : 6</u>.

Treat them just like fractions — use your calculator if you can

Now this is really sneaky. If you stick in a fraction using the [a^b/c] button, your calculator
automatically cancels it down when you press [=].
So for the ratio 8 : 12, just press 8 [a^b/c] 12 [=] , and you'll get the reduced fraction 2/3.
Now you just change it back to ratio form ie. <u>2 : 3</u>. Ace.

The More Awkward Cases:

1) The [a^b/c] button will only accept whole numbers

So <u>IF THE RATIO IS AWKWARD</u> (like "2.4 : 3.6" or "1¼ : 2¾") then you must:
<u>MULTIPLY BOTH SIDES</u> by the <u>SAME NUMBER</u> until they are both <u>WHOLE NUMBERS</u>
and then you can use the [a^b/c] button as before to simplify them down.
e.g. with "1¼ : 2¾", × both sides by 4 gives "<u>5 : 11</u>" (Try [a^b/c], but it won't cancel further)

2) If the ratio is MIXED UNITS

then you must <u>CONVERT BOTH SIDES</u> into the <u>SMALLER UNITS</u> using the
relevant <u>CONVERSION FACTOR</u> (see P.10), and then carry on as normal.
e.g. "36mm : 5.4cm" (× 5.4cm by 10) ⇒ 36mm : 54mm = <u>2:3</u> (using [a^b/c])

3) To reduce a ratio to the form 1 : n   (n can be <u>any number at all</u>)

Simply <u>DIVIDE BOTH SIDES BY THE SMALLEST SIDE</u>.
e.g. take "<u>5 : 27</u>" — dividing both sides by 5 gives: <u>1 : 5.4</u> (27÷5) (i.e. 1 : n)
The 1 : n form is often the <u>most useful</u>, since it shows the ratio very clearly.

Ratios

Using The Formula Triangle in Ratio Questions

"Mortar is made from sand and cement in the ratio 9:2.
If 7 buckets of sand are used, how much cement is needed?"

This is a fairly common type of Exam question and it's pretty tricky for most people
— but once you start using the formula triangle method, it's all a bit of a breeze really.

This is the basic **FORMULA TRIANGLE** for **RATIOS**, but **NOTE**:

$$\frac{A}{A:B \times B}$$

1) **THE RATIO MUST BE THE RIGHT WAY ROUND**,
with the **FIRST NUMBER IN THE RATIO** relating to
the item **ON TOP** in the triangle.

2) You'll always need to **CONVERT THE RATIO** into its
EQUIVALENT FRACTION or Decimal to work out the answer.

The formula triangle for the mortar question is shown below and the trick is to replace
the **RATIO** 9:2 by its **EQUIVALENT FRACTION**: 9/2, or 4.5 as a decimal (9÷2)

So, *covering up cement in the triangle,* gives us "cement = sand / (9:2)"
i.e. "7 / 4.5" = 7 ÷ 4.5 = 1.56 or about *1½ buckets of cement.*

$$\frac{Sand}{9:2 \times Ce...}$$

Proportional Division

In a *proportional division question* a **TOTAL AMOUNT** is to be *split in a certain ratio.*

For example: *"£1800 is to be split in the ratio 2:3:1. Find the 3 amounts".*

The key word here is **PARTS**. — concentrate on "parts" and it all becomes quite painless:

Method

1) **ADD UP THE PARTS:**
The ratio 2:3:1 means there will be a total of 6 *parts*   i.e. 2+3+1 = **6 PARTS**

2) **FIND THE AMOUNT FOR ONE "PART"**
Just *divide* the *total amount* by the number of *parts:*   £1800 ÷ 6 = **£300**   (=1 PART)

3) **HENCE FIND THE THREE AMOUNTS:**
2 parts = 2×300 = **£600**,   3 parts = 3×300 = **£900**,   1 part = **£300**

The Acid Test:
LEARN the **6 RULES** for SIMPLIFYING, the **FORMULA TRIANGLE** for Ratios (plus 2 points to note), and the **3 Steps for PROPORTIONAL DIVISION**.

Now *turn over* and *write down what you've learned.* Try again *until you can do it.*

1) Simplify: a) 20:32   b) 2.6 : 3.9   c) 1¾ : 3½
2) Syrup and ice-cream are mixed in the ratio 2:7 . How much ice-cream should go with
   10 portions of syrup?   3) Divide £5100 in the ratio 9:2:6.

Standard Index Form

Standard Form and Standard Index Form are the SAME THING.
So remember both of these names as well as what it actually is:

Ordinary Number: 5,200,000   In Standard Form: 5.2 X 10^6

Standard form is only really useful for writing VERY BIG or VERY SMALL numbers in a more convenient way, e.g.

37,000,000,000 would be 3.7×10^{10} in standard form.
0.000 000 004 16 would be 4.16×10^{-9} in standard form.

but ANY NUMBER can be written in standard form and you need to know how to do it:

What it Actually is:

A number written in standard form must ALWAYS be in EXACTLY this form:

$$A \times 10^n$$

This number must always
be BETWEEN 1 AND 10.
(The fancy way of saying this is:

"$1 \leqslant A < 10$" — they sometimes
write that in Exam questions — don't let it
put you off, just remember what it means).

This number is just
the NUMBER OF
PLACES
the Decimal Point
moves.

Learn The Three Rules:

1) The front number must always be BETWEEN 1 AND 10

2) The power of 10, n, is purely: HOW FAR THE D.P. MOVES

3) n is +ve for BIG numbers,   n is −ve for SMALL numbers
(This is much better than rules based on which way the D.P. moves.)

Examples:

1) "Express 79 800 in standard form".

METHOD:
1) Move the D.P. until 79 800 becomes 7.98 ("$1 \leqslant A < 10$")
2) The D.P. has moved 4 places so n=4, giving: 10^4
3) 79800 is a BIG number so n is +4, not −4

ANSWER:
7.9800. = 7.98 x 10^4

2) "Express 3.51 x 10^{-3} as an ordinary number".

METHOD:
1) 10^{-3}, tells us that the D.P. must move 3 places...
2) ...and the "−" sign tells us to move the D.P. to make it
a SMALL number. (i.e. 0.00351, rather than 3510)

ANSWER:
3.51 = 0.00351

SECTION ONE — NUMBERS MOSTLY

Standard Index Form

Standard Form and The Calculator

People usually manage all that stuff about moving the decimal point OK *(apart from always forgetting that FOR A BIG NUMBER it's "ten to the power +ve something" and FOR A SMALL NUMBER it's "ten to the power −ve something")*, but when it comes to doing standard form on a *calculator* it's invariably a sorry saga of confusion and ineptitude.
But it's not so bad really — you just have to learn it, that's all.....

1) Entering Standard Form Numbers `EXP`

The button you **MUST USE** to put standard form numbers into the calculator is the `EXP`
(or `EE`) button — but **DON'T** go pressing `X` `10` as well, like a lot of people do,
because that makes it **WRONG**

Example: *"Enter 5.74 x 10⁹ into the calculator"*

Just press: `5.74` `EXP` `9`   and the display will be 5.74^{09}

Note that you **ONLY PRESS** the `EXP` (or `EE`) button — you **DON'T** press `X` or `10` at all.

2) Reading Standard Form Numbers:

The big thing you have to remember when you write any standard form number from the calculator display is to put the "×10" in yourself . **DON'T** just write down what it says on the display.

Example: *"Write down the number* 3.265^{06} *as a finished answer."*

As a finished answer this must be written as 3.265×10^6.

It is **NOT** 3.265^6 so **DON'T** write it down like that — **YOU** have to put the $\times 10^n$ in yourself, even though it isn't shown in the display at all. *That's the bit people forget.*

The Acid Test: LEARN the Three Rules and Two Calculator Methods, then turn over and write them down.

Now cover up these 2 pages and answer these:
1) What are the Three Rules for standard form?
2) Express 927100 in standard index form.   3) And the same for 0.00285
4) Express 7.34×10^4 as an ordinary number.
5) Work this out using your calculator: $2.8 \times 10^{11} \div 4.2 \times 10^{-8}$, and write down the answer, first in standard form and then as an ordinary number.

SECTION ONE — NUMBERS MOSTLY

Powers (or "Indices")

Powers are a very useful shorthand:

$$6\times6 = 6^2 \text{ ("6 squared")}$$
$$2\times2\times2\times2 = 2^4 \text{ ("Two to the power 4")}$$
$$9\times9\times9\times9\times9 = 9^5 \text{ ("Nine to the power 5")}$$
$$4\text{x}4\text{x}4 = 4^3 \text{ ("Four cubed")}$$

That bit is easy to remember. Unfortunately, there are <u>SEVEN SPECIAL RULES</u> for Powers that are not quite so easy, but *<u>you do need to know them for the Exam</u>*:

The Seven Rules

 Powers are ace.

1) When <u>MULTIPLYING</u>, you <u>ADD the powers</u>.

e.g. $2^3 \times 2^4 = 2^{3+4} = 2^7$ $5^2 \times 5^7 = 5^{2+7} = 5^9$

2) When <u>DIVIDING</u>, you <u>SUBTRACT the powers</u>.

e.g. $4^6 \div 4^3 = 4^{6-3} = 4^3$ $10^8/10^3 = 10^{8-3} = 10^5$

3) When <u>RAISING</u> one power to another, you <u>MULTIPLY the powers</u>.

e.g. $(2^2)^4 = 2^{2\times4} = 2^8$, $(5^3)^6 = 5^{18}$

4) <u>$X^1 = X$</u>,   <u>ANYTHING TO THE POWER 1 is just ITSELF</u>

e.g. $3^1 = 3$, $2 \times 2^6 = 2^7$, $5^1 = 5$, $8^3 \div 8^2 = 8^{3-2} = 8^1 = 8$

5) <u>$X^0 = 1$</u>, <u>ANYTHING TO THE POWER 0 is just 1</u>

e.g. $4^0 = 1$ $28^0 = 1$ $2^3 \div 2^3 = 2^{3-3} = 2^0 = 1$

6) <u>$1^x = 1$</u>,   <u>1 TO ANY POWER is still just 1</u>

e.g. $1^{19} = 1$ $1^{63} = 1$ $1^3 = 1$ $1^{202} = 1$

7) <u>FRACTIONAL POWERS means one thing: ROOTS</u>

The Power ½ means *<u>Square Root</u>*,   e.g. $25^{1/2} = \sqrt{25} = 5$

The Power ⅓ means *<u>Cube Root</u>*,   e.g. $27^{1/3} = \sqrt[3]{27} = 3$

The Acid Test:

LEARN the <u>Seven Rules</u> for Powers. Then <u>turn over</u> and <u>write it all down</u>. Keep trying until you can do it!

Then cover the page and then <u>SIMPLIFY</u> these:

1) a) $2^5 \times 2^2$; b) $3^3 \div 3^2$; c) $(7^4)^2$; d) $(4^6 \times 4^2 \times 1^7)$; e) $3^0 \times 1^4 \times 6^2$; f) $(6^3 \times 6^4)/6^0$; g) $8^{10} \div (8 \times 8^3)$.

2) Evaluate these: a) $16^{\frac{1}{2}}$; b) $64^{\frac{1}{2}}$; c) $121^{\frac{1}{2}}$; d) $27^{\frac{1}{3}}$; e) $64^{\frac{1}{3}}$; f) $125^{\frac{1}{3}}$.

Square Roots and Cube Roots

Square Roots

"Squared" means "times by itself" : $P^2 = P \times P$
— SQUARE ROOT is the reverse process.

The best way to think of it is this:

> ### "Square Root" means
> ### "What Number Times by Itself gives..."

Example: "Find the square root of 49" (i.e. " Find $\sqrt{49}$ " or "Find $49^{1/2}$ ")

To do this you should say it as: "What number TIMES BY ITSELF gives... 49"

Now, if you ever felt inclined to learn the number sequences on P.3 (like you were told to!), then of course you'd know instantly that the answer is 7.

> HOVEVER, it has to be said *the best way to find any square root* is simply to use the SQUARE ROOT BUTTON: Press $\boxed{\sqrt{}}$ $\boxed{49}$ $\boxed{=}$ $\underline{7}$ (See P.17)

Cube Roots

"Cubed" means "times by itself three times" : $T^3 = T \times T \times T$
— CUBE ROOT is the reverse process.

> ### "Cube Root" means "What Number
> ### Times by Itself THREE TIMES gives..."

Well, strictly there are only two × signs, but you know what I mean.

Example: "Find the cube root of 64" (i.e. "Find $\sqrt[3]{64}$ " or " Find $64^{\frac{1}{3}}$ ")

You should say: "What TIMES BY ITSELF THREE TIMES gives... 64"

From your in-depth revision of P.3 you will of course know the answer is 4.

> HOVEVER, for others that you may not know, *there's no better method* than to use the CUBE ROOT BUTTON : Press $\boxed{\sqrt[3]{}}$ $\boxed{27}$ $\boxed{=}$ $= \underline{3}$ (See P.17)

And Don't Forget:

"SOMETHING TO THE POWER ½" is just a different way of asking
for a SQUARE ROOT e.g. $81^{1/2}$ is the same as $\sqrt{81}$ which is just $\underline{9}$.

"SOMETHING TO THE POWER 1/3" is just a different way of asking
for a CUBE ROOT e.g. $27^{\frac{1}{3}}$ is the same as $\sqrt[3]{27}$ which is just $\underline{3}$.

The Acid Test:
LEARN the 2 statements in the red boxes, the best method for finding roots and what fractional powers mean.

Now turn over and write down everything you've learned.

1) Use your calculator to find a) $65^{1/2}$ b) $324^{\frac{1}{3}}$ c) $\sqrt{150}$ d) $\sqrt[3]{888}$.
2) a) If $y^2 = 25$, find y. b) If $t^3 = 125$, find t. c) If $10 \times x^3 = 80$, find x.

Revision Summary for Section One

I know these questions seem difficult, _but they're the very best revision you can do_. The whole point of revision, remember, is <u>to find out what you _don't_ know</u> and then learn it <u>until you do</u>. These searching questions test how much you know _better than anything else ever can_. They follow the sequence of pages in Section One, so you can easily look up anything you don't know.

Keep learning these basic facts until you know them

1) What are the multiples of a number?   What are the factors of a number?
2) What is the best method for finding all the factors of a number?
3) What are the prime factors of a number? How do you find them?
4) State the two rules for finding Prime numbers (below 120).
5) List the first ten terms in each of these sequences:
    a) Odd numbers   b) Even numbers   c) Prime numbers
    d) Square Numbers   e) Cube numbers   f) Triangle numbers
6) List the first six equivalent fractions for one fifth.
7) What does FDP stand for? Give full details of the four conversion methods.
8) What are the three steps for rounding off?
9) What are the 3 extra details concerning sig. fig. rounding?
10) State three rules for deciding on appropriate accuracy.
11) State two rules for estimating the answer to a calculation.
12) State two rules for estimating an area or volume.
13) State the 3 steps of the method for applying conversion factors.
14) Give 7 different conversions from one metric unit to another.
15) Give 5 different conversions from one imperial unit to another.
16) Give 8 conversions between metric and imperial units.
17) What does a/b mean?
18) Describe in words the 4 rules for doing fractions by hand.
19) Which is the fraction button?   What must you press to enter 5¼?
20) How would you convert it to a top heavy fraction?
21) Describe the 3 types of percentage question and how to identify them.
22) Give details of the method for each of the 3 types.
23) Give the formula for percentage change, and give 3 examples of it.
24) Explain the difference between the 2 cancel buttons on your calculator.
25) Which are the memory buttons? What are they used for?
26) What does BODMAS mean and what has it got to do with your calculator?
27) When would you use the brackets buttons?
28) Which is the powers button? What must you press to find 7^{12}?
29) Which is the Standard Form button? What must you press to enter 2×10^{-6}?
30) What would the number 8×10^4 look like on the calculator display?
31) How can a calculator help to simplify ratios?
32) What is the formula triangle for ratios?
33) What are the three rules for expressing a number in standard form?
34) What are the seven rules for powers?
35) Explain what a square root is. Explain what a cube root is.

Turning Words into Algebra

This is all about taking a maths problem which is written as a _sentence_ and turning it into an _equation or formula_.   It can seem pretty tricky at first, but _it really isn't that bad_ — once you've learnt how to do it.

Example 1:   "A number is doubled, then three is added to the total, and the result is fifteen. What was the original number?"

Somehow you need to be able to turn that sentence into this equation:

"Solve 2x + 3 = 15"

You have to be the translator:

From long-winded ENGLISH		To ink-saving ALGEBRA
A number	——	x
double it	——	2x
then add three	——	2x + 3
the result is 15	——	2x + 3 = 15

And there you have it. You just have to think about each bit of the _sentence_ and _convert_ it into a bit of _maths_. Keep doing it bit by bit and before you know it you have an equation which you can _solve_. The answer to the above is clearly 6.   ($2 \times 6 + 3 = 15$)

Example 2:   "A gang of four workers were paid £15 per hour plus a tip of £6. They shared the takings and each got £9. How many hours did they work?"

£15 an hour, plus a £6 tip was paid to a gang of four workers. They shared the takings and each got £9.

$$\frac{15x + 6}{4} = 9$$

How many hours did the gang work?

All you have to do now is _solve the equation_. The answer comes out as x = _2 hours_.

Just watch out when translating "LESS":

"**Three less** than a number"   x - 3

means "a number, less three"   NOT 3 - x

The Acid Test: LEARN this page until you can translate all the examples given with the answers covered up.

Basic Algebra

1) Terms

Before you can do anything else, you MUST understand what a TERM is:

1) **A TERM IS A COLLECTION OF NUMBERS, LETTERS AND BRACKETS, ALL MULTIPLIED/DIVIDED TOGETHER.**

2) <u>TERMS are SEPARATED BY + AND − SIGNS</u>   e.g. $5x^2$- 2dy - 3x + 6

3) TERMS always have a   + or −   <u>ATTACHED TO THE FRONT OF THEM</u>

4) e.g.

Invisible
+ sign $8x^2$ + 2xy - 6y + $7y^2$ + 2

"x^2" term   "xy" term   "y" term   "y^2" term   "number" term

2) Simplifying   "Collecting Like Terms"

<u>EXAMPLE:</u>   "Simplify $2x^2 − 5x + 7x^2 + 6x - 3$"

$2x^2$ -5x +$7x^2$ +6x -3   =   +$2x^2$ +$7x^2$ -5x +6x -3

x^2-terms   x-terms   number   = $9x^2$ +x   -3 = $\underline{9x^2 + x - 3}$
term

1) <u>Put bubbles round each term,</u> — be sure you <u>*capture the +/− sign* IN FRONT</u> of each.

2) Then you can <u>*move the bubbles into the best order*</u> so that <u>LIKE TERMS *are together*</u>.

3) <u>"LIKE TERMS"</u> have exactly the same combination of letters, e.g. "x-terms" or "xy-terms".

4) <u>Combine LIKE TERMS</u> using the <u>NUMBER LINE</u> (not the other rule for negative numbers).

3) Multiplying out Brackets

1) The thing <u>OUTSIDE</u> the brackets <u>multiplies each separate term INSIDE the brackets.</u>

2) When letters are <u>multiplied together</u>, they are just <u>written next to each other</u>, pq.

3) Remember, R x R = R^2, and TY^2 means TxYxY, whilst $(TY)^2$ means TxTxYxY.

4) Remember <u>a minus outside the bracket REVERSES ALL THE SIGNS</u> when you multiply.

Examples:

1) $2(4x + 6) = \underline{8x + 12}$

2) $3q(5r − 4q) = \underline{15qr − 12q^2}$

3) $− 5(2a − 3b) = \underline{− 10a + 15b}$ (note a- and b-term signs have been *reversed* — Rule 4)

Basic Algebra

4) Expanding and Simplifying

a) With DOUBLE BRACKETS — you get 4 terms after multiplying

them out and usually 2 of them combine to leave 3 terms, like this:

$$(2A - 5)(3A + 1) = (2A \times 3A) + (2A \times 1) + (-5 \times 3A) + (-5 \times 1)$$
$$= 6A^2 + 2A - 15A - 5$$
$$= \underline{6A^2 - 13A - 5}$$

(these 2 combine together)

b) SQUARED BRACKETS:

e.g. $(5t + 3)^2$ ALWAYS write these out as
two brackets: $(5t + 3)(5t + 3)$ and work them out CAREFULLY like this:

$$(5t + 3)(5t + 3) = 25t^2 + 15t + 15t + 9 = \underline{25t^2 + 30t + 9}$$

YOU SHOULD ALWAYS GET FOUR TERMS from squared brackets, and inevitably *two of these* will combine together to leave THREE TERMS IN THE END, as shown above.

(Also see P. 52)

(The usual WRONG ANSWER, by the way, is $(5t + 3)^2 = 25t^2 + 9$ — eeek!)

5) Factorising — putting brackets in

This is the *exact reverse* of multiplying-out brackets. Here's the method to follow:

1) Take out the biggest NUMBER that goes into all the terms.
2) Take each letter in turn and take out the highest power (e.g. x, x^2 etc.) that will go into EVERY term.
3) Open the brackets and fill in all the bits needed to reproduce each term.

EXAMPLE: Factorise $6x^4y + 8x^2y^3z - 12x^3yz^2$

Answer: $2x^2y(3x^2 + 4y^2z - 6xz^2)$

Biggest number
that'll divide into
6, 8 and 12

Highest powers
of X and Y that will
go into *all three terms*

Z wasn't in ALL terms so
it can't come out as
a *common factor*

REMEMBER:

1) The bits *taken out* and put at the front are the *COMMON FACTORS*.
2) The bits *inside the brackets* are *what's needed to get back to the original terms* if you were to multiply the brackets out again.

The Acid Test:

LEARN the important details for each of the 5 sections on these 2 pages, then turn over and write it all down.

Then apply the methods to these:
1) Simplify: a)   -4x + 2y - 2 - y + x   b)   8w + 6k - 5w - 12k² + 8
2) Expand: a) $2ab(4a - 3b^2)$ b) $(2f+3)(5f-4)$ c) $(1 - 2x)^2$
3) Factorise: a) $12q^2r^3 - 24q^2r + 30q^3r^4$ b) $6x^3y^2 - 3x^2y + 12xy^3z$

Number Patterns

This is an easy topic, but make sure you know <u>ALL SIX</u> types of sequence, not just the first few. The *main secret* is to *write the differences in the gaps* between each pair of numbers. That way you can usually see what's happening whichever type it is.

1) "Common Difference" Type — dead easy

E.g.   4   7   10   13   16   ...   532   537   542   547   552   ...
          +3   +3   +3   +3   +3   +5   +5   +5   +5   +5

2) "Increasing Difference" Type

Here the differences <u>increase by the same amount</u> each time:

E.g.   3   5   8   12   17   23 ...
         +2 +3 +4 +5 +6 +7

3) "Decreasing Difference" Type

Here the differences <u>decrease by the same amount</u> each time:

E.g.   76   66   57   49   42   36   ...
          -10   -9   -8   -7   -6   -5

4) "Multiplying Factor" Type

This type has a common <u>MULTIPLIER</u> linking each pair of numbers:

E.g.   3   6   12   24   48   ...
         ×2   ×2   ×2   ×2   ×2

5) "Dividing Factor" Type

This type has a common <u>DIVIDER</u> linking each pair of numbers:

E.g.   625   125   25   5   ...
            ÷5   ÷5   ÷5   ÷5

6) "Adding Previous Terms" Type

Add the *first two terms* to get the *3rd*, then add the *2nd and 3rd* to get the *4th*, etc.

E.g.   1   1   2   3   5   8   13   21   ....
                1+1 1+2 2+3 3+5 5+8 8+13 13+21

The Acid Test:

LEARN the <u>6 types of number pattern</u>. Then cover the page and answer these:

1) Write down <u>FROM MEMORY</u> the name of each type of number sequence and give an example of each.
2) Find the next two terms in these sequences:
   a) 2,10,50,250...   b) 440,220,110....   c) 2,3,5,8,12...   d) 10,11,13...

SECTION TWO — ALGEBRA

Finding the nth Term

"The nth term" is a formula with "n" in it which gives you every term in a sequence when you put different values for n in. There are two different types of sequence (for "nth term" questions) which have to be done in different ways:

Common Difference Type: "dn + (a – d)"

For any sequence such as   3,   7,   11,   15,   where there's a <u>COMMON DIFFERENCE</u>:

$$\underset{4\quad\quad 4\quad\quad 4}{\nearrow\quad\nearrow\quad\nearrow}$$

you can always find "the nth term" using the <u>FORMULA</u>:   **nth Term = dn + (a–d)**

Don't forget:

> 1) "a" is simply the value of <u>THE FIRST TERM</u> in the sequence.
> 2) "d" is simply the value of <u>THE COMMON DIFFERENCE</u> between the terms.
> 3) To get the <u>nth term</u>, you just <u>find the values of "a" and "d" from the sequence</u>
>   <u>and stick them in the formula</u>.
>   *You don't replace n though — that wants to stay as n*
> 4) <u>Of course YOU HAVE TO LEARN THE FORMULA, but life is like that.</u>

Example: "*Find the nth term of this sequence: 5,   7,   9,   11, ...*"

<u>ANSWER</u>: 1) The formula is dn + (a–d)
2) The <u>first term</u> is 5, so <u>a = 5</u>   The <u>common difference</u> is 2 so <u>d = 2</u>
3) Putting these in the formula gives: nth term = 2n + (5–2)
                                                                      so <u>nth term = 2n + 3</u>

Changing Difference Type:

"a + (n–1)d + ½(n–1)(n–2)C"

If the number sequence is one where the *difference* between the terms is *increasing or decreasing* then it gets a whole lot more complicated (as you'll have spotted from the above formula — which you'll have to *learn!*). This time there are *THREE* letters you have to fill in:

"a" is the <u>FIRST TERM</u>,
"d" is the <u>FIRST DIFFERENCE</u> (between the first two numbers),
"C" is the <u>CHANGE BETWEEN ONE DIFFERENCE AND THE NEXT</u>.

Example: "*Find the nth term of this sequence: 2,   4,   7,   11,*
..."

$$\underset{2\quad\quad 3\quad\quad 4}{\nearrow\quad\nearrow\quad\nearrow}$$

<u>ANSWER</u>:   1) The formula is   "*a + (n–1)d + ½(n–1)(n–2)C*"
2) The <u>first term</u> is 2, so <u>a = 2</u>   The <u>first difference</u> is 2 so <u>d = 2</u>
3) The <u>differences increase</u> by 1 each time so <u>C = +1</u>
Putting these in the formula gives: "*2 + (n–1)2 + ½(n–1)(n–2)×1*"
Which becomes: $2 + 2n - 2 + ½n^2 - 1½n + 1$
Which simplifies to: $½n^2 + ½n + 1$
                                          so the <u>nth term</u> = $½n^2 + ½n + 1$ (Easy peasy, huh!)

The Acid Test: LEARN the <u>definition of the nth term</u> and the <u>4 steps for finding it</u>, and <u>LEARN THE FORMULAE</u>.

1) Find the nth term of the following sequences:
a) 5, 8, 11, 14,.... b) -3, -13, -23,.... c) 1, 3, 6, 10, 15,.... d) 5, 6, 9, 14,...

Negative Numbers and Letters

Everyone knows RULE 1, but sometimes RULE 2 applies instead, so make sure you know *BOTH* rules AND when to use them.

Rule 1

Only to be used when:

+	+	makes	+
+	−	makes	−
−	+	makes	−
−	−	makes	+

1) Multiplying or dividing

e.g. $-1 \times 6 = \underline{-6}$, $-6 \div -3 = \underline{+2}$ $-2y \times -2 = \underline{+4y}$

2) Two signs appear next to each other

e.g. $8 - {}^-7 = 8+7 = \underline{15}$ $3 + {}^-4 - {}^-6 = 3 - 4 + 6 = \underline{5}$

Rule 2

THE NUMBER LINE

Use this when ADDING OR SUBTRACTING:

e.g. "Simplify 7X − 11X − 2X + 4X"

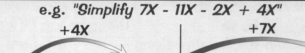

So $7X - 11X - 2X + 4X = \underline{-2X}$

Letters Multiplied Together

This is the super-slick notation they like to use in algebra which just ends up making life difficult for folks like you. You've got to remember these five rules:

1) "abc" means "a×b×c" The ×'s are often left out to make it clearer.

2) "pq²" means "p×q×q" Note that only the q is squared, not the p as well.

3) "(pq)²" means "p×p×q×q" The brackets mean that BOTH letters are squared.

4) "p(q − r)³" means "p×(q − r) × (q − r) × (q − r)" Only the brackets get cubed.

5) "−2²" is too ambiguous. It should either be written $(-2)^2 = 4$, or $-(2^2) = -4$.

The Acid Test:

LEARN the Two Rules for negative numbers and the cases where each of them is used and the 5 special cases of letters multiplied together.

Then turn over and write down what you've learned.

1) For each of a) to d), decide where Rule 1 and Rule 2 apply, and then work them out.
  a) -7×-2; b) $-2 + {}^-3 + 1$; c) $(2X + {}^-3X - 5X) \div (3 + {}^-9)$; d) $10 \div {}^-2$.

2) If x=3 and y=−2, work out: a) xy^2; b) $(xy)^3$; c) $x(2+y)^2$; d) $x^3 + y^2 + 2x^2y^3$.

SECTION TWO — ALGEBRA

Substituting Values into Formulae

This topic is a lot easier than you think

$$C = \frac{5}{9}(F - 32)$$

Generally speaking, algebra is a pretty grim subject, but you should realise that some bits of it are VERY easy, and this is <u>definitely the easiest bit of all</u>, so whatever you do, <u>don't pass up on these easy Exam marks</u>.

Method

If you don't follow this STRICT METHOD you'll just keep getting them wrong — it's as simple as that.

1) <u>Write out the Formula</u>

    e.g. $C = \frac{5}{9}(F - 32)$

2) <u>Write it again</u>, directly underneath, but <u>substituting numbers for letters</u> on the RHS.
  (Right Hand Side)

    $C = \frac{5}{9}(80 - 32)$

3) Work it out <u>IN STAGES</u>.

   Use <u>BODMAS</u> to work things out <u>IN THE RIGHT ORDER</u>.
   <u>WRITE DOWN</u> values for each bit <u>as you go along</u>.

    $C = \frac{5}{9}(48)$

    $= 240 \div 9$

    $\underline{C = 26.7°}$

4) <u>DO NOT</u> attempt to do it <u>all in one go</u> on your calculator.
   That ridiculous method <u>fails at least 50% of the time</u>!

BODMAS

<u>B</u>rackets, <u>O</u>ther, <u>D</u>ivision, <u>M</u>ultiplication, <u>A</u>ddition, <u>S</u>ubtraction

BODMAS tells you the ORDER in which these operations should be done: Work out <u>brackets</u> first, then <u>Other</u> things like squaring, then <u>multiply</u> / <u>divide</u> groups of numbers before <u>adding</u> or <u>subtracting</u> them. This set of rules works really well for simple cases, so remember the word: BODMAS.   (See P.18)

Example

A mysterious quantity Q, is given by: $Q = (R - 5)^2 + 7T/W$
Find the value of Q when R = 8, T = 3 and W = -7

<u>ANSWER</u>:
1) Write down the formula: $Q = (R - 5)^2 + 7T/W$
2) Put the numbers in: $Q = (8 - 5)^2 + 7 \times 3/-7$
3) Then work it out <u>in stages</u> : $= (3)^2 + 21/-7$
                           $= 9 + -3$
                           $= 9 - 3 \ = \underline{6}$

<u>Note BODMAS in operation</u>:

<u>Brackets</u> worked out first, then <u>squared</u>. <u>Multiplications</u> and <u>divisions</u> done <u>before</u> finally <u>adding</u> and <u>subtracting</u>.

The Acid Test:

<u>LEARN</u> the <u>4 Steps of the Substitution Method</u> and the <u>full meaning of BODMAS</u>. Then turn over.....

... and write it all down from memory.   1) Practise the above example until you can do it easily without help.   2) If $C = \frac{5}{9}(F - 32)$, find the value of C when F = 68.

Solving Equations The Easy Way

The "proper" way to solve equations is shown on P.37. In practise the "proper way" can be pretty difficult so there's a lot to be said for the much easier methods shown below.

The drawback with these is that you can't always use them on very complicated equations. In most Exam questions though, they do just fine.

1) THE "COMMON SENSE" APPROACH

The trick here is to realise that the unknown quantity "X" is after all just a number and the "equation" is just a cryptic clue to help you find it

Example: *"Solve this equation: 2X + 3 = 27"*

                 (i.e. find what number **X** is)

Answer: *This is what you should say to yourself:*

> "<u>Something + 3 = 27</u>"   hmm, so that "something" must be 24.
>
> So that means 2X = 24, which means "2 times something = 24"
>
> So it must be 24 ÷ 2 which is 12   so <u>X = 12</u> "

In other words don't think of it as algebra, but as "<u>Find the mystery number</u>".

2) THE TRIAL AND ERROR METHOD

This is a perfectly good method, and although it won't work every time, it usually does, especially if the answer is a <u>whole number</u>.

The *big secret of trial and error* methods is to <u>find TWO OPPOSITE CASES</u> and <u>keep taking values IN BETWEEN</u> them.

In other words, find a number that makes the <u>RHS bigger</u>, and then one that makes the <u>LHS bigger</u>, and then try values *in between them*. (See P.35)

Example: *"Solve for X: 4X + 2 = 20 – 5X"*

                 (i.e. find the number X)

Answer:

> Try X=1:   4+2 = 20 – 5,   6 = 15 — no good, <u>RHS too big</u>
>
> Try X=3:   12 + 2 = 20 – 15,   14 = 5 — no good, <u>now LHS too big</u>

<u>SO TRY IN BETWEEN</u>:   X = 2:   8 + 2 = 20 – 10,   10 = 10,   YES, so <u>X = 2</u>.

The Acid Test: LEARN these two methods until you can <u>turn the page and write them down</u> with an example for each.

1) Solve: 3x – 8 = 25   2) Solve: 2x + 4 = 6x – 4

Trial and Improvement

In principle, this is an easy way to find approximate answers to quite complicated equations, especially "cubics" (ones with x^3 in). BUT... you have to make an effort to <u>LEARN THE FINER DETAILS</u> of this method, otherwise you'll never get the hang of it.

Method

1) <u>SUBSTITUTE TWO INITIAL VALUES</u> into the equation that give <u>OPPOSITE CASES</u>. These are usually suggested in the question. If not, you'll have to think of your own. Opposite cases means <u>one answer too big, one too small</u>, or <u>one +ve, one –ve</u>, for example. If they don't give opposite cases <u>try again</u>.

2) Now CHOOSE YOUR NEXT VALUE <u>IN BETWEEN</u> THE PREVIOUS TWO, and <u>SUBSTITUTE it into the equation</u>.
   <u>Continue this process</u>, always choosing a new value <u>between the two closest opposite cases</u>, (and preferably nearer to the one which is closest to the answer you want).

3) <u>AFTER ONLY 3 OR 4 STEPS</u> you should have <u>2 numbers</u> which are to the <u>right degree of accuracy</u> but DIFFER BY 1 IN THE LAST DIGIT.
   For example if you had to get your answer to 2 DP then you'd eventually end up with say 5.43 and 5.44, with these giving OPPOSITE results of course.

4) <u>At this point</u> you ALWAYS take the <u>Exact Middle Value</u> to decide which is the answer you want.   e.g. for 7.45 and 7.46, you'd try 7.455 to see if the real answer was <u>between 7.45 and 7.455</u> or between <u>7.455 and 7.46</u> (see below).

Example

"The equation $X^3 + X = 16$ has a solution between 2 and 2.5. Find this solution to 1 DP"

Try X = 2	$2^3 + 2 = 10$	(Too small)	← (2 opposite cases)
Try X = 2.5	$2.5^3 + 2.5 = 18.125$	(Too big)	

16 is what we want and it's closer to 18.125 than it is to 10 so we'll choose our next value for X closer to 2.5 than 2.

Try X = 2.3	$2.3^3 + 2.3 = 14.467$	(Too small)

Getting close, next we need to see if 2.4 is still too big or too small:

Try X = 2.4	$2.4^3 + 2.4 = 16.224$	(Too big)

Good, now we know that <u>the answer must be between 2.3 and 2.4</u>. To find out which one it's nearest to, we have to try the EXACT MIDDLE VALUE: 2.35

Try X = 2.35	$2.35^3 + 2.35 = 15.328$	(Too small)

This tells us with certainty that the solution must be between 2.35 (too small) and 2.4 (too big), and so to 1 DP <u>it must round up to 2.4</u>. ANSWER = 2.4

The Acid Test:

"LEARN and TURN" — if you don't actually <u>commit it to memory</u>, then you've wasted your time even reading it.

To succeed with this method you must <u>LEARN the 4 steps above</u>. Do it now, and practise until you can <u>write them down without having to look back at them</u>.
It's not as difficult as you think.

1) The equation $X^3 - X = 3$ has a solution between 1 and 2. Find it to 1 DP.

The Balance Method for Equations

Or, "How to peel off the wrapping from your X" ...

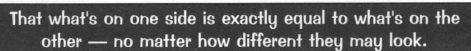

But not too seriously!

Get it - "eagle" sign...

You Must Take the "=" Sign Seriously

When you see an "=" sign you must realise what it actually means:

> That what's on one side is exactly equal to what's on the other — no matter how different they may look.

That means you're allowed to do _anything you like_ to one side — so long as you do _exactly the same thing_ to the other. That's really important, so don't forget it.

Anything Divided by Itself = 1

You ought to know what's really going on when, for example, you change an equation

from $5x = 18$ to $x = \dfrac{18}{5}$

The big trick here is to remember that: $\dfrac{ANYTHING}{ITSELF} = 1$

EXAMPLE: Solve the equation $5x = 18$.

ANSWER: First you should _divide by 5_ on both sides like this: $\dfrac{5x}{5} = \dfrac{18}{5}$ (This is what you would always do: divide both sides by what the x had been multiplied by — in this case, 5.)

The _really important_ bit to get into your head is _this_: $\dfrac{5x}{5} = \left(\dfrac{5}{5}\right)x = 1x = x$

So that means we end up with: $x = \dfrac{18}{5} = 3.6$

Peeling off the + and − Terms

$2x - y = 7$

$2x - \underbrace{y + y}_{zero} = 7 + y$

$2x \qquad = 7 + y$

$x \quad = \quad \dfrac{7 + y}{2}$

Compare these "before" and "after" shots and you'll see the _reason_ why the rule _"change the sign when crossing from one side of the = to the other"_ actually works.

$2x + y = 7$

$2x + \underbrace{y - y}_{zero} = 7 - y$

$2x \qquad = 7 - y$

$x \quad = \quad \dfrac{7 - y}{2}$

The Acid Test:

LEARN the **3 SECTIONS** on this page. Then turn over and scribble down everything you've learned.

This page has the "proper" explanations on it, specially for those of you who want to know the reasons for things. Just to make sure you've got it firmly fastened in there, do this question:

1) Write out a step-by-step solution for $5(3x - 2) = 35$. Use colours as in the above example.

Solving Equations

Solving Equations means finding the value of x from something like: 6x + 7 = 1 − 4x.
Now, not a lot of people know this, but *exactly the same method applies* to both
solving equations and   *rearranging formulas*, as illustrated on these two pages.

1) EXACTLY THE SAME METHOD APPLIES TO BOTH FORMULAS AND EQUATIONS.
2) THE SAME SEQUENCE OF STEPS APPLIES EVERY TIME.

To illustrate the sequence of steps we'll use this equation: $\sqrt{3 - \dfrac{2x + 2}{x + 4}} = 2$

The Six Steps Applied to Equations

1) Get rid of any square root signs by <u>squaring both sides</u>: $3 - \dfrac{2x + 2}{x + 4} = 4$

2) Get everything off the bottom by <u>cross-multiplying up to EVERY OTHER TERM</u>:

$$3 - \frac{2x + 2}{x + 4} = 4 \quad \Rightarrow \quad 3(x + 4) - (2x + 2) = 4(x + 4)$$

3) Multiply out any brackets: $3x + 12 - 2x - 2 = 4x + 16$

**4) Collect all <u>subject terms</u> on one side of the "=" and all <u>non-subject</u>
terms on the other side, <u>remembering to reverse the +/− sign of any</u>
<u>term that crosses the "="</u> :**

+4x moves across the "=" and becomes -4x
+12 moves across the "=" and becomes -12
-2 moves across the "=" and becomes +2

$$3x - 2x - 4x = 16 - 12 + 2$$

**5) <u>Combine together like terms</u> on each side of the equation, and reduce it
to the form "<u>Ax = B</u>", where A and B are just numbers (or bunches of
letters in the case of formulas):**

$$-3x = 6$$
("Ax = B" :   A = -3, B = 6, x is the subject)

6) Finally <u>slide the A underneath the B</u> to give "X = $\frac{B}{A}$".
divide, and that's your answer:

$$x = \frac{6}{-3} = -2 \quad \text{So } \underline{x = -2}$$

The Acid Test:
LEARN the <u>6 STEPS</u> for <u>solving equations</u> and
<u>rearranging formulas</u>. Turn over and write them down.

1) Solve the following equations: a) 3(x + 1) = 2 + 4(2 − x)   b) $\dfrac{6}{x + 3} = \dfrac{9}{5 + 2x}$

Rearranging Formulas

Rearranging Formulas means making one letter the subject, e.g. getting "y= " from something like $3x + z = 5(y + 4w)$.
Generally speaking "solving equations" is easier, but don't forget:

1) EXACTLY THE SAME METHOD APPLIES TO BOTH FORMULAS AND EQUATIONS.
2) THE SAME SEQUENCE OF STEPS APPLIES EVERY TIME.

We'll illustrate this by making "y" the subject of this formula: $H = \sqrt{4G - \dfrac{E^2}{2y+1}}$

The Six Steps Applied to Formulas

1) Get rid of any square root signs by <u>squaring both sides</u>: $H^2 = 4G - \dfrac{E^2}{2y+1}$

2) Get everything off the bottom by
   <u>cross-multiplying up to EVERY OTHER TERM</u>:

$$H^2 = 4G - \frac{E^2}{2y+1} \implies H^2(2y + 1) = 4G(2y + 1) - E^2$$

3) Multiply out any brackets: $\qquad 2yH^2 + H^2 = 8Gy + 4G - E^2$

4) Collect all <u>subject terms</u> on one side of the "=" and all <u>non-subject</u> <u>terms</u> on the other side, <u>remembering to reverse the +/− sign of any</u> <u>term that crosses the "="</u> :

+8Gy moves across the "=" and becomes −8Gy
+H² moves across the "=" and becomes −H²

$$2yH^2 - 8Gy \ = \ -H^2 + 4G - E^2$$

5) <u>Combine together like terms</u> on each side of the equation, and reduce it to the form "<u>Ax = B</u>", where A and B are just bunches of letters which DON'T include the subject (y). Note that the LHS has to be <u>FACTORISED</u>:

$$(2H^2 - 8G)y = 4G - H^2 - E^2$$
("Ax = B" i.e. $A = (2H^2 - 8G)$, $B = 4G - H^2 - E^2$, y is the subject)

6) Finally <u>slide the A underneath the B</u> to give "$X = \frac{B}{A}$".
   (cancel if possible) and that's your answer:

$$\text{So} \quad y = \frac{4G - H^2 - E^2}{2H^2 - 8G}$$

The Acid Test:
LEARN the <u>6 STEPS</u> for <u>solving equations</u> and <u>rearranging formulas</u>. Turn over and write them down.

1) Rearrange " $F = \frac{9}{5}C + 32$ " from "F= ", to "C= " and then back the other way.
2) Make m the subject of these: a) $\dfrac{m}{m+n} = 7$; b) $\dfrac{1}{m} = \dfrac{1}{n} + \dfrac{1}{p}$.

Formula Triangles

You may have already come across these in physics, but whether you have or you haven't, the fact remains that they're *extremely potent tools* for quite a number of tricky maths problems — so make sure you know how to use them.
They're *very easy to use* and *very easy to remember*. Watch:

If 3 things are related by a formula that looks either

like this: $A = B \times C$ or like this: $B = \dfrac{A}{C}$

then you can put them into a FORMULA TRIANGLE like this:

1) First decide where the letters go:

1) If there are <u>TWO LETTERS MULTIPLIED TOGETHER</u> in the formula then they must go <u>ON THE BOTTOM</u> of the Formula Triangle (and so *the other one* must go *on the top*).

For example the formula "<u>F = m×a</u>" fits into
a formula triangle like this →

2) If there's <u>ONE THING DIVIDED BY ANOTHER</u> in the formula then the one <u>ON TOP OF THE DIVISION</u> goes <u>ON TOP IN THE FORMULA TRIANGLE</u> (and so the other two must go *on the bottom* — it doesn't matter which way round).

For example the formula " <u>SINθ = Opp/Hyp</u>" fits into a formula triangle like this ↑ .

2) Using the Formula Triangle:

Once you've got the formula triangle sorted out, the rest is easy:

1) <u>COVER UP</u> *the thing you want to find* and just <u>WRITE DOWN</u> *what's left showing*.
2) <u>PUT IN THE VALUES</u> for the other two things and just <u>WORK IT OUT</u>.

Example:

"Using " <u>F = m×a</u>*" , find the value of "a" when F = 15 and m = 50"*

<u>ANSWER</u>: Using the formula triangle, we want to find "a" so we cover "a" up, and that leaves "F/m" showing (i.e. F÷m).
So "a = F/m", and putting the numbers in we get: a = 15/50 = <u>0.3</u>.

The Acid Test:
<u>LEARN THIS WHOLE PAGE</u> then turn over and <u>write down</u> all the important details including the examples.

Density and Speed

You might think this is physics, but density is specifically mentioned in the maths syllabus, and it's very likely to come up in your Exam. The standard formula for density is:

Density = Mass ÷ Volume

so we can put it in a <u>FORMULA TRIANGLE</u> like this:

One way or another <u>you MUST remember this formula for density</u>, because they won't give it to you and without it you'll be pretty stuck. <u>The best method by far</u> is to <u>remember the order of the letters</u> in the FORMULA TRIANGLE as D^MV or <u>DiMoV</u> (The Russian Agent!).

EXAMPLE:

"Find the volume of an object which has a mass of 60g and a density of 2.4g/cm³"

<u>ANSWER</u>: To find volume, <u>cover up V</u> in the formula triangle. This leaves M/D showing, so V = M ÷ D

$$= 60 \div 2.4$$
$$= \underline{25 \text{ cm}^3}$$

Speed = Distance ÷ Time

This is very common. In fact it probably comes up every single year — *and they never give you the formula!* Either <u>learn it beforehand</u> or wave goodbye to <u>lots of easy marks</u>. Life isn't all bad though — there's an easy <u>FORMULA TRIANGLE</u>:

Of course you still have to <u>remember the order of the letters</u> in the triangle (SDT) — but this time we have the word <u>SoDiT</u> to help you.

So if it's a question on speed, distance and time just say: **SOD IT**.

EXAMPLE:

"A car travels 140 miles at 40 miles per hour. How long does it take?"

ANSWER: <u>We want to find the TIME</u>, so <u>cover up T</u> in the triangle which leaves D/S,

so T = D/S = Distance ÷ speed = 140÷40 = <u>3.5 hours</u>

> **LEARN THE <u>FORMULA TRIANGLE</u>, AND YOU'LL FIND QUESTIONS ON *SPEED, DISTANCE AND TIME* <u>VERY EASY</u>.**

The Acid Test:

LEARN the formulas for <u>DENSITY</u> and <u>SPEED</u> — and also the two <u>Formula triangles</u>.

1) What's the formula triangle for Density?
2) A metal object has a volume of 53cm³ and a mass of 656g. What is its density?
3) Another piece of the same metal has a volume of 46cm³. What is its mass?
4) What's the formula for speed, distance and time?
5) Find the time taken, for a person walking at 1.8 km/h to cover 45km.
   Also, find how far she'll walk in 10 hrs 30 mins.

Two Hints When Using Formulas

These are just the kind of little details that you really do need to know but somehow never quite get to learn — *WELL LEARN THEM NOW!*

1) Units — Getting them Right

By *units* we mean things like *cm, m, m/s, km²* etc. and as a rule you don't have to worry too much about them. However, when you're using a FORMULA TRIANGLE, there's one special thing you need to know. It's simple enough *but you must know it*:

The UNITS you get OUT of a Formula DEPEND ENTIRELY upon the UNITS you put INTO IT

So for example if you put a *distance in CM* and a *time in SECONDS* into the formula triangle to work out speed, the answer must come out in *CM per SECOND* (cm/s).

Alternatively, if the *time is in HOURS* and the *speed in MILES PER HOUR* (mph) then the *distance* you'd calculate would obviously come out as MILES.

It's pretty simple when you think about it. Where you really have to watch out is when you get this sort of question:

Example
"A boy walks 600m in 15 minutes. Find his speed in km/h"

ANSWER: If you just do *"600m ÷ 15 minutes"* your answer will be a speed, sure, but in metres per minute (m/min) which is no good at all.

Instead you must CONVERT INTO KM AND HOURS first:
600m = 0.6 km   15 mins = 0.25 hours (mins÷60).
Then you can divide 0.6 *km* by 0.25 *hours* to get 2.4 km/h which is much more like it.

2) Converting Time to Hrs, Mins and Secs with ⊙′″

Here's a tricky detail that comes up when you're doing speed, distance and time: converting an answer like 1.45 HOURS into HOURS AND MINUTES.

What it DEFINITELY ISN'T is 1 hours and 45 mins — remember, your calculator DOES NOT work in hours and minutes UNLESS YOU TELL IT TO, like this:

1) ENTERING a time in hours, mins and seconds:

e.g. to enter 5hrs 23mins and 42 secs, press 5 ⊙′″ 23 ⊙′″ 42 ⊙′″ =. The display shows the time as 5.395 hours, which is *the correct number for using in a formula because it's a DECIMAL*. NEVER enter 5hrs 23mins as [5.23] — it's horrific.

2) Converting a decimal time into hours, mins and secs:

To convert 1.45 hours (which is what your calculator will give you) into hrs, mins and secs you press 1.45 =, and then SHIFT ⊙′″ and the display will then show 1° 27° 0, which means 1 hour, 27 mins (and 0 secs), which is just what you need for the final answer.

The Acid Test:
LEARN the two important topics on this page, then turn over and write down everything you've learned.

1) Find the time taken, in *hours, mins and secs*, to travel 4,352 m at a speed of 8 km/h.

SECTION TWO — ALGEBRA

X, Y and Z Coordinates

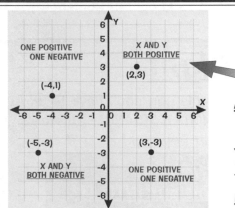

A graph has <u>four different regions</u> where the X- and Y- coordinates are either <u>positive</u> or <u>negative</u>.

This is the easiest region by far because here <u>ALL THE COORDINATES ARE POSITIVE.</u>

You have to be *dead careful* in the *OTHER REGIONS* though, because the X- and Y- coordinates could be <u>negative</u>, and that always makes life difficult.

X, Y Coordinates — getting them in the right order

You must always give <u>COORDINATES</u> in brackets like this: (x,y)

$$(\ x\ ,\ y\ )$$

And you always have to be real careful to get them *the right way round* , **X** first, then **Y**. Here are *THREE POINTS* to help you remember:

1) The two coordinates are always in <u>ALPHABETICAL ORDER, X then Y</u>.

2) X is always the flat axis going <u>ACROSS</u> the page.
In other words " <u>X is a..cross</u> "   Get it! - x is a "×".   (Hilarious isn't it)

3) Remember it's always <u>IN THE HOUSE</u> (→) and then <u>UP THE STAIRS</u> (↑), so it's <u>ALONG first</u> and <u>then UP</u>, i.e. X-coordinate first, and then Y-coordinate.

3-D Coordinates — a Very Easy Topic

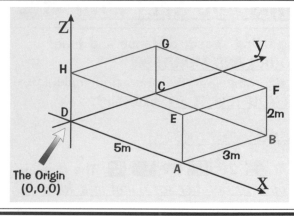

1) All it involves is *extending* the normal x-y coordinates into a *third direction*, z, so that *all positions then have 3 coordinates: (x,y,z)*.

2) This means you can give the coordinates of the *corners of a box* or any other <u>3-D SHAPE</u>.

For example in this drawing, the coordinates of A and B are   A(5,0,0)   B(5,3,0)

The Acid Test:

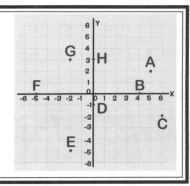

> LEARN the <u>3 Rules for getting X and Y the right way round</u>. Then turn over and <u>write it all down</u>.

1) Write down the coordinates of the letters **A** to **H** on this graph:
2) Write down the coordinates of all the remaining corners of the box shown above.

Easy Graphs You Should Know

If you want to make life easy for yourself, then you <u>*definitely*</u> need to know a few simple graphs straight off <u>*without even having to blink*</u>.   These are they:

1) "<u>X = a</u>"
<u>VERTICAL</u> Lines

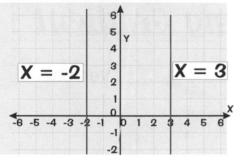

<u>"X = a number"</u>   is a line that goes <u>straight up through that number</u> on the X-axis,   e.g. X = 3 goes straight up through 3 on the X-axis as shown.
Don't forget: <u>the y-axis is also the line "x = 0"</u>

2) "<u>Y = a</u>"
<u>HORIZONTAL</u> Lines

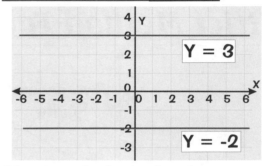

<u>"Y = a number"</u>   is a line that goes <u>straight across through that number</u> on the Y-axis,   e.g. Y = -2 goes straight through -2 on the Y-axis as shown.
Don't forget: <u>the x-axis is also the line "y = 0"</u>

3) "<u>Y = X</u>" and "<u>Y = −X</u>"
(The Main Diagonals)

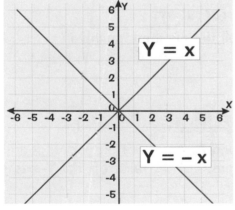

"<u>Y = X</u>" is the <u>main diagonal</u> that goes <u>UPHILL</u> from left to right.

"<u>Y = -X</u>" is the <u>main diagonal</u> that goes <u>DOWNHILL</u> from left to right.

4) "<u>Y = AX</u>" and "<u>Y = −AX</u>"
(Other Sloping Lines)

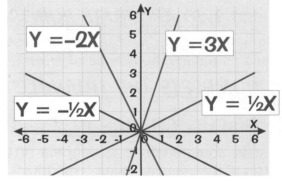

<u>Y = AX</u> and <u>Y = -AX</u> are the equations for <u>A SLOPING LINE THROUGH THE ORIGIN</u>.

The value of A is <u>the *GRADIENT* of the line</u>, so <u>the BIGGER the number the STEEPER the slope</u>, and a MINUS SIGN tells you it slopes DOWNHILL as shown by the ones above.

The Acid Test: LEARN the <u>FOUR EASY TYPES OF GRAPH</u>, then <u>turn over</u> and <u>WRITE IT ALL DOWN</u> with examples.

Then <u>cover the page</u> and do these:
1) Write down the equations of <u>the four graphs shown here:</u>
2) Draw these 6 graphs: X = 2,   Y = -1, Y = X,   Y = -X, X = 0,   Y = -3X.

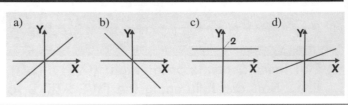

Finding the Gradient of a Line

Working out the gradient of a straight line is a slightly involved business, and there are quite a few things that can go wrong.

Once again though, if you _learn and follow the steps below_ and treat it as a <u>STRICT METHOD</u>, you'll have a lot more success than if you try and fudge your way through it, like you usually do.

Strict Method For Finding Gradient

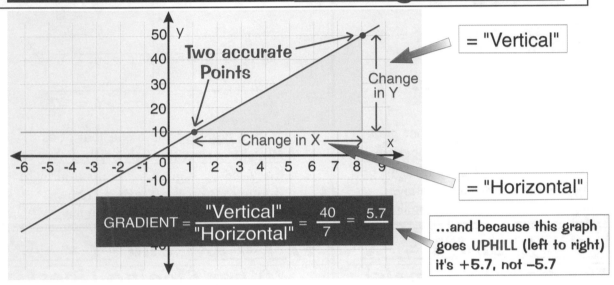

1) Find _TWO ACCURATE POINTS_, reasonably far apart

Both in the _upper right quadrant_ if possible (to keep all the numbers positive and so reduce the chance of errors). (See P.42)

2) _COMPLETE THE TRIANGLE_ as shown

3) Find the <u>CHANGE IN Y</u> and the <u>CHANGE IN X</u>

Make sure you do this _using the SCALES on the Y- and X- axes_, <u>not by counting cm</u>! (So in the example shown, the Change in Y is NOT 4cm, but _40 units_ off the Y-axis.)

4) _LEARN_ this formula, and use it:

$$\text{GRADIENT} = \frac{\text{VERTICAL}}{\text{HORIZONTAL}}$$

Make sure you get it the right way up too! Remember it's
<u>VER</u>y <u>HO</u>t — <u>VER</u>tical over <u>HO</u>rizontal.

5) Finally, is the gradient _POSITIVE_ or _NEGATIVE_?

If it slopes <u>UPHILL</u> left → right (⟋) <u>then it's +ve.</u>
If it slopes <u>DOWNHILL</u> left → right (⟍) <u>then it's −ve.</u> (so put a minus(−) in front of it)

The Acid Test:

<u>LEARN</u> the <u>FIVE STEPS</u> for finding a gradient then <u>turn over</u> and <u>WRITE THEM DOWN</u> from memory.

1) Plot these 2 points on a graph: (0,4) (1,0) and then join them up with a straight line. Now carefully apply the <u>FIVE STEPS</u> to find the gradient of the line.

What the Gradient of a Graph Means

1) In Real Life

No matter what the graph, <u>THE MEANING OF THE GRADIENT</u> is always simply :

(Y-axis UNITS) PER (X-axis UNITS)

EXAMPLES:

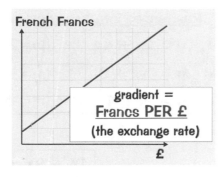

French Francs

gradient =
<u>Francs PER £</u>
(the exchange rate)

£

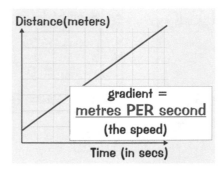

Distance(meters)

gradient =
<u>metres PER second</u>
(the speed)

Time (in secs)

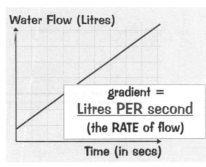

Water Flow (Litres)

gradient =
<u>Litres PER second</u>
(the RATE of flow)

Time (in secs)

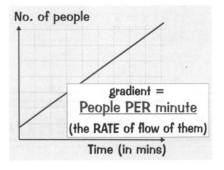

No. of people

gradient =
<u>People PER minute</u>
(the RATE of flow of them)

Time (in mins)

Some gradients have special names like *Exchange Rate* or *Speed*, but once you've written down *"something PER something"* using the Y-axis and X-axis <u>UNITS</u>, it's then pretty easy to work out what the gradient represents.

2) In the Equation of the Line

(See P. 43)

Here you see the 2x family of slopes.

Each section of them goes
<u>up 2</u> for every <u>1 across</u>.

The slope, or gradient, <u>is whatever the x is multiplied by</u>, in this case 2.

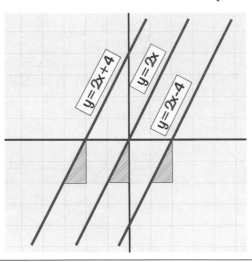

y=2x+4

y=2x

y=2x-4

Four Graphs You Should Recognise

There are four types of graph that you should know the basic shape of just from looking at their equations — it really isn't as difficult as it sounds.

1) Straight Line Graphs: "Y = mx + c"

— which means "*Y = gradient-times-X + place where it crosses Y axis*".

Straight line equations are really quite easy to spot — they have an *x-term*, a *y-term* and *a number* and that's it. There's no x^2 or x^3 or $\frac{1}{x}$ terms or any other fancy things.

NOT straight lines	Straight lines		Rearranged into "y = mx +c"
$x^3 = 2 - y$	$y = 4 + 2x$	$\rightarrow \quad y = 2x + 4$	(m=2, c=4)
$y = x^2 + 2$	$x - 3y = 0$	$\rightarrow \quad y = \frac{1}{3}x + 0$	(m=⅓, c=0)
$1/y + 2/x = 5$	$2y + 2x = 8$	$\rightarrow \quad y = -x + 4$	(m=-1, c=4)

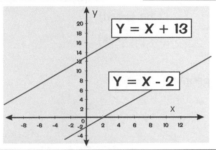

Y = X + 13

Y = X - 2

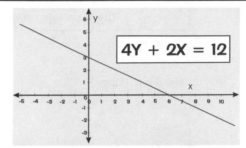

4Y + 2X = 12

2) X^2 Bucket Shapes:

Y = anything with X^2 in it, but not X^3

Notice that all these X^2 graphs have the same **SYMMETRICAL** bucket shape.

Also notice that if the X^2 bit is positive (i.e. $+X^2$) then the bucket is the normal way up, but if the x^2 bit has a "minus" in front of it (i.e. $-X^2$) then the bucket is <u>upside down</u>.

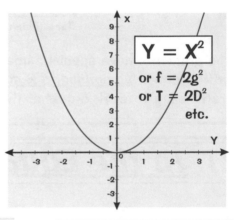

$Y = X^2$

or $f = 2g^2$
or $T = 2D^2$
etc.

Did someone say bucket shaped giraffe?

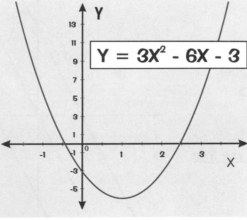

$Y = 3X^2 - 6X - 3$

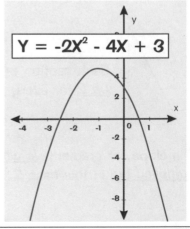

$Y = -2X^2 - 4X + 3$

Four Graphs You Should Recognise

3) $\underline{X^3}$ *Graphs:*

$\underline{Y =}$ "something with X^3 in it"

All X^3 graphs have the same basic *wiggle* in the middle, but it can be a flat wiggle or a more pronounced wiggle.

Notice that "$\underline{-X^3}$ graphs" always come *down from top left* whereas the $\underline{+X^3}$ ones go *up from bottom left*.

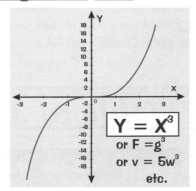

$$Y = X^3$$
or $F = g^3$
or $v = 5w^3$
etc.

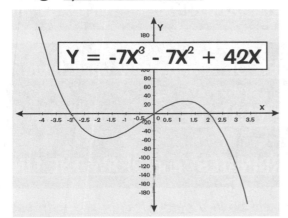

$$Y = -7X^3 - 7X^2 + 42X$$

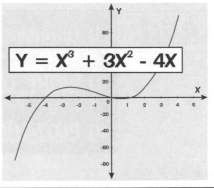

$$Y = X^3 + 3X^2 - 4X$$

4) *1/X GRAPHS:*

$\underline{Y = }^A\!/_X$, where A is some number.

These graphs are all EXACTLY the same shape, the only difference being how close in they get at the corner. They are all *symmetrical about the line y=x*. This is also the graph you get when x and y are in *inverse proportion*.

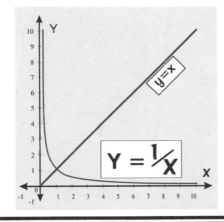

$$Y = {}^1\!/_X$$

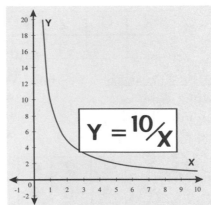

$$Y = {}^{10}\!/_X$$

The Acid Test:

LEARN all the details about the 4 Types of Graph, their equations and their shapes.

Then *turn over* and *sketch three examples* of each of the *four types* of graph — and if you can also give some extra details about their equations, *so much the better*.
Remember, if you don't LEARN IT, then it's a waste of time even reading it. This is true for all revision.

Plotting Straight Line Graphs

A lot of people wouldn't know a straight line equation if it ran up and bit them, but they're pretty easy to spot really — they just have _two letters_ and _a few numbers_, but with _nothing fancy_ like squared or cubed. (Now that you're inflamed with burning curiosity, look at P.46 to see some examples)

Anyway, in your GCSE you'll be expected to draw the graph of a straight line equation. "Y = mX + c" is the hard way of doing it (see P.49). Here's the __EASY WAY__ of doing it:

The "Table of 3 values" method

You can __EASILY__ draw the graph of __ANY EQUATION__ using this __EASY__ method.

Method:

1) Choose __3 VALUES OF X__ and __draw up a table__,

2) __WORK OUT THE VALUE OF Y__ for each value of X.

3) __PLOT THE COORDINATES__, and __DRAW THE LINE__.

If it's a straight line equation, the 3 points will be in a dead straight line with each other, which is the usual check you do when you've drawn it.
If they aren't, then it could be a curve and you'll need to do more values in your table to find out what's going on.

Example: _"Draw the graph of Y = 2X – 4"._

1) __DRAW UP A TABLE__ with some _suitable values_ of X.
   Choosing X = 0, 2, 4 is usually cool enough.
   i.e.

X	0	2	4
Y			

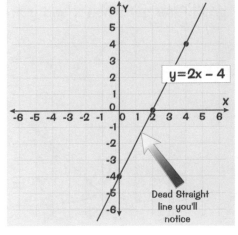

y=2x – 4

Dead Straight
line you'll
notice

2) __FIND THE Y-VALUES__   e.g. When $\underline{X = 4}$,
   by putting each $y = 2X – 4$
   x-value into the $= 2 \times 4 – 4$
   equation: $= 8 – 4$ $= \underline{4}$

X	0	2	4
Y	-4	0	4

3) __PLOT THE POINTS__ and __DRAW THE LINE__ right across the graph (as shown).

   (The points should always lie in a __DEAD STRAIGHT LINE__. If they don't, do more values in the table to find out what on earth's happening.)

The Acid Test: __LEARN__ the details of this _easy method_ then _turn over and write them all down_.

1) Draw the graphs of   a) y = 2x - 3;   b) y = 5 + x;   c) y = 5 – 2x.

SECTION TWO — ALGEBRA

Plotting Straight Line Graphs

Using y = mx + c

y = mx +c is the general equation for a straight line graph, and you need to remember:

"m" is equal to the <u>GRADIENT</u> of the graph
"c" is the value <u>WHERE IT CROSSES THE Y-AXIS</u> and is called the <u>INTERCEPT</u>.

1) Drawing a Straight Line using "y = mx + c"

The main thing is being able to identify "m" and "c" and knowing what to do with them:

BUT WATCH OUT — people often mix up "m" and "c", especially with say, "y = 5 + 2x"
<u>REMEMBER</u>: "m" is the number <u>IN FRONT OF X</u> and "c" is the number <u>ON ITS OWN</u>.

Method

1) Get the equation into the form "<u>y = mx + c</u>".

2) <u>IDENTIFY</u> "m" and "c" *CAREFULLY*.

3) <u>PUT A DOT ON THE Y-AXIS</u> at the value of c.

4) Then go <u>ALONG ONE UNIT</u> and <u>up or down by the value of m</u> and make another dot.

5) <u>Repeat</u> the same "step" in <u>both directions</u> as shown:

6) Finally <u>CHECK</u> that the gradient <u>LOOKS RIGHT</u>.

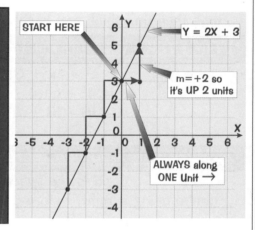

The graph shown here shows the process for the equation "y = 2x + 3":

1) "c" = 3, so put a first dot at y = 3 on the y-axis.
2) Go along 1 unit → and then up by 2 because "m" = +2.
3) Repeat the same step, 1→ 2↑ in <u>both</u> directions. (i.e. 1 ← 2 ↓ the other way)
4) CHECK: <u>a gradient of +2</u> should be <u>quite steep and uphill left to right</u> — which it is.

2) Finding the Equation Of a Straight Line Graph

<u>THIS IS EASY</u>:
1) Find where the graph <u>CROSSES THE Y-AXIS</u>.
 This is the value of "c".
2) Find the value of the <u>GRADIENT</u> (see P.44).
 This is the value of "m"
3) Now just put these values for "m" and "c" into "y = mx + c" and there you have it!

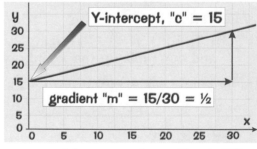

For the graph shown here, m=½ and c = 15 so "y = mx + c" becomes "<u>y = ½x + 15</u>"

The Acid Test:

<u>LEARN THE DETAILS</u> of the two methods for "y = mx +c".
Then <u>TURN OVER</u> and <u>WRITE DOWN</u> what you've learned.

1) Using "y = mx +c" draw the graphs of y=x − 1 and y=3 − 2x.
2) Using "y = mx +c" find the equations of these 3 graphs →

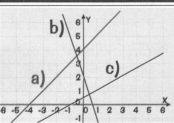

SECTION TWO — ALGEBRA

Typical Graph Questions

There's a lot of fiddly details involved in graph questions: getting the right values in the table; plotting the right points; and getting the final answers from your graph.
If you want to get all these easy marks, then you've got to learn all these little tricks:

Filling in The Table of Values

A typical question: *"Complete the table of values for the equation $y = x^2 – 4x + 3$"*

x	-2	-1	0	1	2	3	4	5	6
y				0			3		15

<u>WHAT YOU DON'T DO</u> is try to punch it all into the calculator in one go. Not good.
The rest of the question hinges on this table of values and one silly mistake here could cost you a lot of marks. This might look like a long-winded method but it takes far less time than you think and is the only <u>REALLY SAFE</u> method.

1) For EVERY value in the table you should WRITE THIS OUT:

<u>For x=4</u>: $y = x^2 – 4x + 3$
$= 4^2 – 4{\times}4 + 3$
$= 16 – 16 + 3$
$= \underline{3}$

<u>For x=-1</u>: $y = x^2 – 4x + 3$
$= (-1{\times}-1) – (4{\times}-1) + 3$
$= 1 – -4 + 3 = 1 + 4 + 3$
$= \underline{8}$

2) Make sure you can reproduce the y-values they've already given you...

— *BEFORE you fill in the spaces in the table.*   This is really important to make sure you're doing it right, before you start cheerfully working out a pile of wrong values!

I wouldn't tell you all this without good reason, so ignore it at your peril.

Plotting the Points and Drawing the Curve

Here again there are easy marks to be won and lost — this all matters:

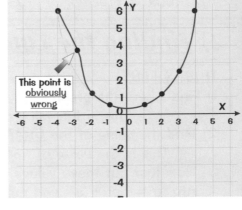

This point is <u>obviously wrong</u>

1) <u>GET THE AXES THE RIGHT WAY ROUND</u>: The values from the <u>FIRST</u> row or column are ALWAYS plotted *on the X-axis.*

2) <u>PLOT THE POINTS CAREFULLY</u>, and don't mix up the x and y values.

3) The points will ALWAYS form a <u>DEAD STRAIGHT LINE</u> or a <u>COMPLETELY SMOOTH CURVE</u>. If they don't, they're *wrong*.

4) A graph from an <u>ALGEBRA EQUATION</u>, must always be drawn as a <u>SMOOTH CURVE</u> (or a dead straight line). You only use lots of short straight line sections to join points in *"Data Handling"* when it's called a "frequency polygon". (See P.88)

<u>NEVER EVER</u> *let one point drag your line off* in some ridiculous direction — if one point seems out of place, *check the value in the table* and then check the position where you've plotted it. When a graph is generated from an equation, *you never get spikes or lumps* — only MISTAKES.

Typical Graph Questions

Getting Answers from Your Graph

1) <u>FOR A SINGLE CURVE OR LINE</u>, you <u>ALWAYS</u> get the answer by *drawing a straight line to the graph from one axis*, and then <u>down or across to the other axis</u>, as shown here:

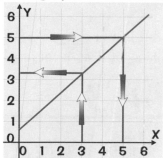

You should be *fully expecting* this to happen so that even if you don't understand the question, you can still have a pretty good stab at it:

If the question said *"Find the value of y when x is equal to 3"*, <u>ALL YOU DO IS THIS</u>: start at *3* on the x-axis, go straight up to the graph, then straight over to the y-axis and read off the value, which in this case is <u>*y = 3.2*</u> (as shown opposite).

2) <u>IF TWO LINES CROSS.....</u>

you can bet your very last fruitcake the answer to one of the questions will simply be:

<u>THE VALUES OF X AND Y WHERE THEY CROSS</u> and you should be expecting that *before they even ask it!* (See Simultaneous Eqns. P.54).

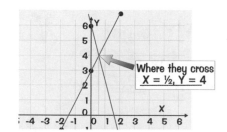

Where they cross
X = ½, Y = 4

Travel Graphs

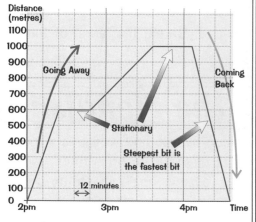

THE FOUR KEY POINTS:

1) A <u>TRAVEL GRAPH</u> is always <u>DISTANCE</u> (↑) against <u>TIME</u> (→).
2) For any section, <u>SLOPE (gradient) = SPEED</u>, but watch out for the <u>UNITS</u>.
3) <u>FLAT SECTIONS</u> are where it's <u>STOPPED</u>.
4) The <u>STEEPER</u> the graph the <u>FASTER</u> it's going.

A Typical Tricky Question:

"What's the speed of the return section on the graph shown above?"

<u>ANSWER</u>: Speed = gradient = 1000m/30mins = 33.33 <u>m/min</u> (metres per minute)

or   1km ÷ ½hr = <u>2 km/h</u> (kilometres per hour)

or   1000m ÷ 1800s = <u>0.56 m/s</u> (metres per second)

Note that the <u>answer</u> (and its units) depends very much on what units you use to <u>work it out</u>.

The Acid Test:

LEARN the <u>2 Rules for doing tables of values</u>, the <u>4 points for drawing graphs</u> and the <u>2 simple Rules for getting answers</u>.

Now turn over and *write it all down from memory*. Then *try again until you can do it*.

1) *Complete the table of values* at the top of the previous page (using the proper methods!), and then *draw the graph* taking note of the Four Points.
2) From your graph <u>find the value of y when x is 4.2</u>, and <u>the values of x when y=12</u>.
3) If I drew a graph of "miles covered" up the y-axis and "gallons used" along the x-axis, and worked out the gradient, what would the value of it tell me?
4) For the travel graph shown above, work out the speed of the middle section, giving your answer in km/h. Also, describe what is happening between 2pm and 4:36pm.

SECTION TWO — ALGEBRA

Expanding Out Brackets

"Expanding out brackets" just means _"multiplying brackets together"_,   e.g. (x + 3)(x + 2).
This can get quite tricky and it's _all too easy_ to get it wrong.
Learn these _two very handy methods_ and always use one or the other:

The Jolly old "Area" Method

It's pretty cool to imagine the result of _multiplying two brackets_ a bit like an _area_.
This is done by looking at something like (x + 3)(x + 2), and thinking of it as a rectangle
which is "(x + 3) long" by "(x + 2) wide".
This is shown here:

X   **+ 3**

MULTIPLY to get
each of the _4 BITS_,
...and then _ADD_ the
4 bits together...

X

This bit is _x long_ by _x wide_: AREA x^2	This bit is _3 long_ by _x wide_: AREA 3x
This bit is _x long_ by _2 wide_: AREA 2x	This bit is _3 long_ by _2 wide_: AREA 6

+2

... and you get the "area" of the entire (x + 3)(x + 2) rectangle:

$$x^2 + 3x + 2x + 6 = \underline{x^2 + 5x + 6}$$

The Fiendish Foil Method

The other method, which is a more "_grown up_" method, is to just _multiply_ each of the
four bits straight off without using the area idea or drawing any boxes.
For reasons which should be _quickly obvious_ we call this the _FOIL method_.

Firsts: (x + 3) (x + 2)

Outsides: (x + 3) (x + 2)

Insides: (x + 3) (x + 2)

Lasts: (x + 3) (x + 2)

So we end up with

F + O + I + L =

$x^2 + 2x + 3x + 6 =$

$\underline{x^2 + 5x + 6}$ (again)

This is the better method to use.

The Acid Test:

LEARN both the _area_ and _foil_ methods of expanding.
Then turn over and scribble them down.

Then choose one of them and expand the following:
1) (x + 1)(x + 7);   2) (x - 1)(x + 3);   3) (x + 6)(x - 2);   4) (x - 4)(x - 5).

SECTION TWO — ALGEBRA

Factorising Quadratics

What it Means

"Factorising a quadratic" means *"putting it back into 2 brackets"* — pity they don't just call it *"contracting"*, because it *is* the exact opposite of *"expanding"*. There are several different methods for doing it, so stick with the one you're happiest with. If you have no preference then learn this one. The standard format for any quadratic equations is:

$$ax^2 + bx + c = 0$$ (e.g. $x^2 + 5x + 3 = 0$)

Factorising Method

1) **ALWAYS** rearrange into the **STANDARD FORMAT**: $ax^2 + bx + c = 0$.

2) Write down the **TWO BRACKETS** with the x's in: $(x\quad)(x\quad)=0$.

3) Then <u>find 2 numbers</u> that **MULTIPLY to give "c"** (the end number) but also **ADD/SUBTRACT to give "b"** (the coefficient of x).

4) Put them in and check that the +/− signs work out properly.

Example

"Solve $x^2 - x = 6$ by factorising."

<u>ANSWER</u>: 1) <u>First rearrange it</u> (into the standard format): $x^2 - x - 6 = 0$

2) The initial brackets are (as ever): $(x\quad)(x\quad)=0$

3) We now want to look at <u>all pairs of numbers</u> that <u>multiply to give "c"</u> (=6), but which also <u>add or subtract to give the value of b</u>: (-1)

1×6	*Add/subtract to give:*	7 or 5
2×3	*Add/subtract to give:*	5 or ① ← this is what we're after (= ±b)

4) So 2 and 3 will give $b = \pm 1$, so put them in: $(x\quad 2)(x\quad 3)=0$

5) <u>Now fill in the +/− signs</u> so that the 2 and 3 add/subtract to give -1 (=b), Clearly it must be +2 and − 3 so we'll have: $(x + 2)(x - 3)=0$

6) <u>As an ESSENTIAL check, EXPAND the brackets</u> out again to make sure they give the original equation:
$(x + 2)(x - 3)=\ x^2 + 2x - 3x - 6\ =\ x^2 - x - 6$

<u>We're not finished yet mind</u>, because $(x + 2)(x - 3)=0$ is only the <u>factorised form</u> <u>of the equation</u> — we have yet to give the actual **SOLUTIONS**. This is very easy:

7) **THE SOLUTIONS** are simply <u>the two numbers in the brackets</u>, but with **OPPOSITE +/− SIGNS**: i.e.   <u>x = -2 or +3</u>

Make sure you remember that last step. <u>It's the difference</u> between <u>SOLVING THE EQUATION</u> and merely <u>factorising it</u>.

The Acid Test: LEARN the <u>7 steps</u> for solving quadratics by <u>factorising</u>.

1) Solve these <u>by the factor method</u>: a) $x^2 + 2x - 8 = 0$ b) $x^2 + 5x - 24 = 0$
   c) $x^2 - x - 12 = 0$ d) $x^2 + 3x - 20 = 8$

Simultaneous Equations

These are OK as long as you learn these <u>SIX STEPS</u> in every meticulous detail.

The Six Steps

We'll use these two equations for our example:
$$y + 6x = -1 \quad \text{and} \quad 2y = 13 + 3x$$

1) REARRANGE BOTH EQUATIONS INTO THE FORM: $ax + by = c$

where a,b,c are numbers,   (which can be negative).
Also <u>LABEL THE TWO EQUATIONS</u> —① and —②

$$6x + y = -1 \qquad —①$$
$$-3x + 2y = 13 \qquad —②$$

2) You need to <u>MATCH UP THE NUMBERS IN FRONT</u> (the "coefficients")

of either the x's or y's in <u>BOTH EQUATIONS</u>.
To do this you may need to <u>MULTIPLY</u> one or both equations by a suitable number. You should then <u>RELABEL</u> them: —③ and —④

$$6x + y = -1 \qquad —③$$
$$②\times 2: \quad -6x + 4y = 26 \qquad —④$$

(This gives us +6x in equation —③ to match the –6x in equation —②, now called —④)

3) <u>ADD OR SUBTRACT THE TWO EQUATIONS</u> ...

...to eliminate the terms with the same coefficient.
If the <u>coefficients are the SAME</u> (both +ve or both –ve) then <u>SUBTRACT</u>.
If the <u>coefficients are OPPOSITE</u> (one +ve and one –ve) then <u>ADD</u>.

$$③+④ \quad 0x + 5y = 25$$

(In this case we have +6x and –6x so we ADD)

4) <u>SOLVE THE RESULTING EQUATION</u> to find whichever letter is left in it.

$$5y = 25 \Rightarrow \underline{y = 5}$$

5) <u>SUB THIS BACK</u> into equation ① and solve it to find the other quantity.

Sub in ①: $6x + 5 = -1 \Rightarrow 6x = -6 \Rightarrow \underline{x = -1}$

6) Then <u>SUBSTITUTE BOTH THESE VALUES INTO EQUATION</u> ② to make

sure it works out properly. If it doesn't then you've done something wrong and you'll have to do it all again!

Sub x and y in ②: $-3 \times -1 + 2 \times 5 = 3 + 10 = \underline{13}$

which is right, so it's worked.

So the solutions are: $\underline{x = -1}$, $\underline{y = 5}$

The Acid Test:

LEARN the *6 Steps* for solving *Simultaneous Equations*.

Remember, you only know them when you can write them all out from memory, so turn over the page and try it. Then apply the 6 steps to find V and W given that
$$3V - 2 = -4W \quad \text{and} \quad W = 4V - 28$$

Simultaneous Equations

On the opposite page is the *tricky algebra method* for solving simultaneous equations.
On this page is the *nice easy graph method* for solving them.
You could be asked to do *either* method in the Exam so make sure you *learn them both*.

Solving Simultaneous Equations Using Graphs

This is a very easy way to find the x- and y- solutions to two simultaneous equations.
Here's the simple rule:

> **THE SOLUTION** OF TWO **SIMULTANEOUS EQUATIONS** IS SIMPLY THE **X** AND **Y** VALUES **WHERE THEIR GRAPHS CROSS**

Three Step Method

1) Do a *"TABLE OF 3 VALUES"* for both equations.

2) Draw the Two *GRAPHS*.

3) Find the X- and Y-values *WHERE THEY CROSS*.

Easy Peasy.

Example

"Draw the graphs for "Y = 2X + 1" and "Y = 4 – X" and then use your graphs to solve them."

1) **TABLE OF 3 VALUES**
   for both equations:

$y = 2X + 1$

X	0	1	2
Y	1	3	5

$y = 4 - x$

X	0	2	4
Y	4	2	0

2) **DRAW THE GRAPHS**

3) **WHERE THEY CROSS**,
   $x = 1, y = 3$.
   And that's the answer!

   $\underline{x = 1 \text{ and } y = 3}$

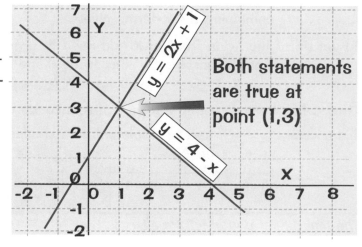

Both statements are true at point (1,3)

The Acid Test:

LEARN the Simple fact that the solution is where the lines cross.

Then try the graphical method on the following:

$$y = 3x + 1; \quad y = -2x - 2$$

Inequalities

This is basically quite difficult, but it's still worth learning the easy bits in case they ask a very easy question on it, as well they might. Here are the easy bits:

The 4 Inequality Symbols:

> means "<u>Greater than</u>" $\geq$ means "<u>Greater than or equal to</u>"

< means "<u>Less than</u>" $\leq$ means "<u>Less than or equal to</u>"

<u>REMEMBER</u>, the one at the <u>BIG</u> end is <u>BIGGEST</u>

so "X > 3" and "3 < X" <u>BOTH</u> say: "<u>X is greater than 3</u>"

Algebra With Inequalities — this is generally a bit tricky

The thing to remember here is that <u>inequalities are just like regular equations</u>:

$$6X > X + 4$$
$$6X = X + 4$$

in the sense that <u>all the normal rules of algebra</u> (See P.28) <u>apply</u> ....

...<u>BUT WITH ONE BIG EXCEPTION</u>:

Whenever you MULTIPLY OR DIVIDE BY A <u>NEGATIVE NUMBER</u>, you must <u>FLIP THE INEQUALITY SIGN</u>.

Example: "<u>Solve 3X < 4X + 3</u>"

ANS: First move the 4X over the "=" : $3X - 4X < 3$

combining the X-terms gives: $-X < 3$

To get rid of the "−" in front of X you need to <u>divide both sides by -1</u> — but remember that means the "<" has to be flipped as well, which gives:

$X > -3$ i.e. "<u>X is greater than -3</u>" is the answer

(The < has flipped around into a >, because we divided by a −ve number)

This answer, <u>X > −3</u>, can be displayed as a shaded region on a number line like this:

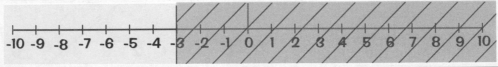

<u>The main thing you should realise</u>, is that <u>MOST OF THE TIME you just treat the "<" or ">" as though it was an "="</u> and <u>do all the usual algebra</u> that you would for a regular equation. The "<u>Big Exception</u>" doesn't actually come up very often at all.

The Acid Test:
LEARN: The <u>4 Inequality Signs</u>, the <u>similarity</u> with <u>EQUATIONS</u> and the <u>One Big Exception</u>.

Now <u>cover the page and write down what you've learned</u>.

1) Solve this inequality: $5X - 2 \leq 6X + 2$.

2) Find all the integer values of X which satisfy both $5X < 3 + 4X$ and $3X + 7 \geq 1$

Graphical Inequalities

This is easy so long as you remember the easy method for drawing graphs — i.e. a table of 3 values (see P.48).

The questions always involve <u>SHADING A REGION ON A GRAPH</u>, which is actually easy but it's always presented as some really horrid-looking algebra that puts most people right off before they even start.

The thing is, once you realise that the horrid-looking algebra just means something really simple then the whole thing becomes quite mind-numbingly simple.   (!)

Method

1) <u>CONVERT</u> each <u>INEQUALITY</u> to an <u>EQUATION</u>

by simply putting an "=" in place of the "<"

2) <u>DO A TABLE OF 3 VALUES FOR EACH EQUATION</u> (See P.48)

and then <u>draw the lines</u> on the graph.

3) <u>SHADE THE ENCLOSED REGION</u>

The lines you've drawn will always enclose the region that you're after — and they nearly always ask you to <u>shade it</u>.

Example

"Shade the region represented by : $y > 2$, $y < x + 1$ and $x + y < 7$"

(See what I mean about the horrid-looking algebra)

<u>ANSWER</u>:

1) <u>CONVERT EACH INEQUALITY TO AN *EQUATION*</u>:

$y > 2$ becomes $y = 2$

$y < x + 1$ becomes $y = x + 1$,

$x + y < 7$ becomes $x + y = 7$.

2) <u>DO A TABLE OF 3 VALUES</u> for each equation, and draw the lines on a graph.

e.g. for $x + y = 7$:

X	0	3	6
Y	7	4	1

3) <u>SHADE THE ENCLOSED REGION</u>,

and Bob's your Uncle, it's done.

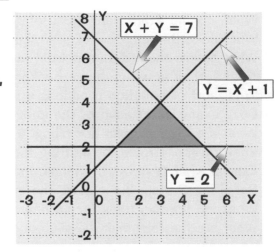

The Acid Test:

LEARN the <u>Three Steps</u> for doing <u>graphical inequalities</u>, then <u>turn over</u> and <u>write them down</u>.

1) Show on a graph the region enclosed by the following three conditions:

$y < 2x$, $x + y < 5$, $y > 1$.

Revision Summary for Section Two

I know these questions seem difficult, but they are the very best revision you can do. The whole point of revision, remember, is to find out what you don't know and then learn it until you do. These searching questions test how much you know better than anything else ever can. They follow the sequence of pages in Section Two, so you can easily look up anything you don't know.

Keep learning these basic facts until you know them

1) Write down three examples of translating a question from English into algebra.
2) In algebra, what is a term? What are the four steps for simplifying an expression?
3) Give the four most important details to do with multiplying out brackets.
4) What happens with double and squared brackets? Give the 3-step method for factorising.
5) Give all six different types of number pattern with an example of each.
6) What are the two formulae for finding the nth term of a number pattern?
7) What are the two rules for negative numbers? When are they used?
8) List 5 combinations of letters that cause confusion in algebra, e.g. ab².
9) What are the 4 steps of the method for substituting numbers into formulas?
10) What has BODMAS got to do with putting numbers into formulas?
11) What are the two easier alternative methods for tackling simple equations?
12) Demonstrate your prowess at these methods by doing an example of each.
13) Give the 4 steps for solving an equation by trial and improvement.
14) Explain what the "balance method" is for solving equations.
15) What do solving equations and rearranging formulae have in common?
16) List the 6 step method for doing equations and formulae. What is meant by Ax = B?
17) What are the two steps for using a formula triangle?
18) What is the formula triangle for   a) density?   b) speed?
19) What do you know about the units that come out of a formula?
20) Explain what 3-D co-ordinates are. Give an example.
21) What type of line is   a) X=a;   b) Y=b;   c) Y=AX.   Sketch Y=X and Y=-X.
22) What is the formula for gradient? How do you remember it?
23) Write down the 5 step method for finding gradient.
24) What are the 4 different types of graph that you should know the basic shape of?
25) What does the equation of a straight line look like?
26) What makes them different from equations that aren't straight lines?
27) What sort of equation has a bucket shaped graph? What about an upside-down bucket?
28) What sort of equation produces a graph with a wiggle in the middle?
29) Do 2 examples of each of the above three, giving the equation and sketching the graph.
30) What are the two rules for filling in a table of values?
31) What are the 4 rules for plotting the graph from a table of values?
32) Explain what "Y=mX+c" means, including the significance of "m" and "c".
33) Detail the 3 steps for getting the equation of a straight line graph.
34) What are the two rules for getting answers from a graph or graphs?
35) What are the four main details concerning travel graphs?
36) What does "factorising a quadratic" mean you have to do?
37) What are "simultaneous equations"? Give an example.
38) Give the 6 step method for solving simultaneous equations.
39) Detail 3 steps of the method for solving simultaneous equations using graphs.
40) What are the four inequality symbols and what do they mean?
41) What are the rules of algebra for inequalities? What is the big exception?
42) What are the 3 steps for doing graphical inequalities?

SECTION TWO — ALGEBRA

Regular Polygons

A <u>POLYGON</u> is a <u>MANY-SIDED SHAPE</u>. A <u>REGULAR</u> polygon is one where <u>ALL THE SIDES AND ANGLES ARE THE SAME</u>. The <u>REGULAR POLYGONS</u> are a <u>never-ending</u> series of shapes with some fancy features. <u>They're very easy to learn</u>. Here are the first few but they don't stop — you can have one with 16 sides or 30, etc.

EQUILATERAL TRIANGLE

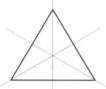

<u>3 sides</u>
<u>3 lines</u> of symmetry
Rot^{nl} symm. <u>order 3</u>

SQUARE

<u>4 sides</u>
<u>4 lines</u> of symmetry
Rot^{nl} symm. <u>order 4</u>

REGULAR PENTAGON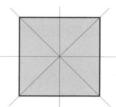

<u>5 sides</u>
<u>5 lines</u> of symmetry
Rot^{nl} symm. <u>order 5</u>

REGULAR HEXAGON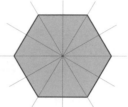

<u>6 sides</u>
<u>6 lines</u> of symmetry
Rot^{nl} symm. <u>order 6</u>

REGULAR HEPTAGON

<u>7 sides</u>
<u>7 lines</u> of symmetry
Rot^{nl} symm. <u>order 7</u>

A 50p piece is like a heptagon

REGULAR OCTAGON

<u>8 sides</u>
<u>8 lines</u> of symmetry
Rot^{nl} symm. <u>order 8</u>

Interior And Exterior Angles

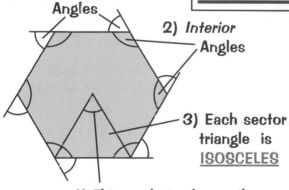

1) *Exterior Angles*

2) *Interior Angles*

3) Each sector triangle is <u>ISOSCELES</u>

4) This angle is always the same as the Exterior Angles

This is the <u>MAIN BUSINESS</u>. Whenever you get a <u>Regular Polygon</u>, it's a <u>cosmic certainty</u> you'll need to work out the <u>Interior and Exterior Angles</u>, because they are the <u>KEY</u> to it all.

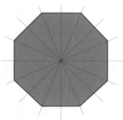

EXTERIOR ANGLE $= \dfrac{360^{0}}{\text{No. of Sides}}$

INTERIOR ANGLE $=$ $180^{0} -$ EXTERIOR ANGLE

The Acid Test:

LEARN EVERYTHING ON THIS PAGE.
Then cover it up and answer these little jokers:

1) What is a Regular Polygon?   2) Name the first six of them.
3) Draw a Pentagon and a Hexagon and put in all their lines of symmetry.
4) What are the two important formulae to do with polygons?
5) Work out the two key angles for a Hexagon.   6) And for a 10-sided Regular Polygon.

Symmetry

SYMMETRY is where a shape or picture can be put in DIFFERENT POSITIONS that LOOK EXACTLY THE SAME. There are THREE TYPES of symmetry:

1) Line Symmetry

This is where you can draw a MIRROR LINE (or more than one) across a picture and both sides will fold exactly together.

| 1 LINE OF SYMMETRY | 2 LINES OF SYMMETRY | 1 LINE OF SYMMETRY | 3 LINES OF SYMMETRY | 1 LINE OF SYMMETRY | NO LINES OF SYMMETRY |

How to draw a reflection:

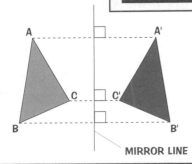

MIRROR LINE

1) Reflect each point one by one

2) Use a line which crosses the mirror line at 90° and goes EXACTLY the same distance on the other side of the mirror line, as shown.

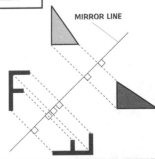

MIRROR LINE

2) Plane Symmetry

Plane Symmetry is all to do with 3-D SOLIDS. Whereas flat shapes can have a mirror line, solid 3-D objects can have planes of symmetry.

A plane mirror surface can be drawn through many regular solids, but the shape must be EXACTLY THE SAME ON BOTH SIDES OF THE PLANE (i.e. mirror images), like these are:

Planes of Symmetry

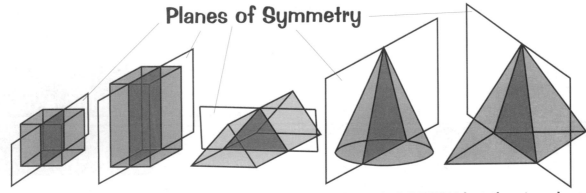

The shapes drawn here all have MANY MORE PLANES OF SYMMETRY but there's only one drawn in for each shape, because otherwise it would all get really messy and you wouldn't be able to see anything.

Symmetry

3) Rotational Symmetry

This is where you can <u>ROTATE</u> the shape or drawing into different positions that <u>all</u> <u>look exactly the same.</u>

Order 1

Order 2

Order 2

Order 3

Order 4

Two Key Points:

1) The <u>ORDER OF ROTATIONAL SYMMETRY</u> is the fancy way of saying:
<u>"HOW MANY DIFFERENT POSITIONS LOOK THE SAME"</u>.
e.g. you should say the Z shape above has "<u>Rotational symmetry order 2</u>"

2) BUT...when a shape has <u>ONLY 1 POSITION</u> you can <u>EITHER</u> say that it has
"<u>Rotational Symmetry order 1</u>"   <u>OR</u>   that it has "<u>NO Rotational Symmetry</u>"

Tracing Paper

SYMMETRY IS ALWAYS A LOT EASIER WITH TRACING PAPER.

1) For <u>REFLECTIONS</u>, trace one side of the drawing and the mirror line too. Then
<u>TURN THE PAPER OVER and line up the mirror line</u> in its original position again.
(If you put a blob on the mirror line it helps you get it back in position again)

2) For <u>ROTATIONS</u>, just swizzle the tracing paper round.
It's really good for <u>finding the CENTRE of rotation</u> (by trial and error)
as well as the <u>order of rotational symmetry</u>.

3) You can use tracing paper in the <u>EXAM</u> —   so <u>ASK FOR IT</u>, or else take your
own in with you.

The Acid Test:

<u>LEARN</u> the important details about <u>LINE AND PLANE</u> <u>SYMMETRY</u>, the <u>2</u> points about <u>ROTATIONAL</u> <u>SYMMETRY</u> and the <u>3</u> points about <u>TRACING PAPER</u>.

Now <u>TURN OVER</u> and <u>WRITE IT ALL DOWN</u> *with examples*, to see what you've learned.

1) Copy these letters and mark in all the <u>lines of symmetry</u>.
Also say what the <u>rotational symmetry</u> is for each one.

I   N   E   Y   W   S   T   Z

2) Copy all the five solids on the last page <u>without their plane of symmetry</u>
(see P.62). Then draw in a <u>different</u> plane of symmetry for each one.
(Drawing 3-D objects ain't easy but it's good laughing at everyone else's dismal efforts.)

The Shapes You Need To Know

These are easy marks in the Exam — make sure you know them all.

1) SQUARE

4 lines of symmetry.
Rotational symmetry order 4

2) RECTANGLE

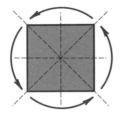
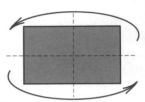

2 lines of symmetry.
Rotational symmetry order 2

3) RHOMBUS (A square pushed over)
(It's also a diamond)

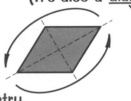
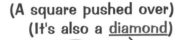

2 lines of symmetry.
Rotational symmetry order 2

4) PARALLELOGRAM
(A rectangle pushed over — two pairs of parallel sides... see p68)

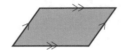

NO lines of symmetry.
Rotational symmetry order 2

5) TRAPEZIUM (One pair of parallel sides)

Only the Isosceles trapezium has a line of symmetry.
None have rotational symmetry

6) KITE

1 line of symmetry.
No rotational symmetry

7) EQUILATERAL Triangle

60, 60, 60

3 lines of symmetry.
Rotational symmetry order 3

8) RIGHT-ANGLED Triangle

No symmetry unless the angles are 45°

9) ISOSCELES Triangle

2 sides equal
2 angles equal

1 line of symmetry.
No rotational symmetry

10) SOLIDS

 REGULAR TETRAHEDRON

 CUBE

 CYLINDER

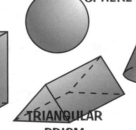 CUBOID

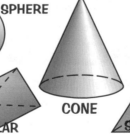 SPHERE

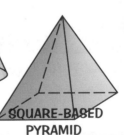 CONE

TRIANGULAR PRISM

SQUARE-BASED PYRAMID

The Acid Test: LEARN everything on this page.

Then turn over and write down all the details that you can remember. Then try again.

Areas

YES IT'S TRUE, these formulae are given inside the front cover of the Exam, but I GUARANTEE that if you don't learn them beforehand, you'll be *totally incapable* of using them in the Exam – *REMEMBER, I ABSOLUTELY GUARANTEE IT* !

YOU MUST LEARN THESE FORMULAE:

1) RECTANGLE

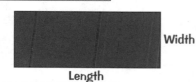

Area of *RECTANGLE* = length × width

$$A = l \times w$$

2) TRIANGLE

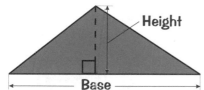

Area of *TRIANGLE* = ½ × base x vertical height

$$A = \tfrac{1}{2} \times b \times h_v$$

Note that the *height* must always be the *vertical height*, not the sloping height.

3) PARALLELOGRAM

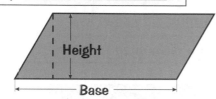

Area of *PARALLELOGRAM* = base × vertical height

$$A = b \times h_v$$

4) TRAPEZIUM

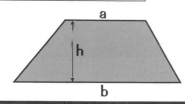

Area of *TRAPEZIUM* = average of parallel sides × distance between them

$$A = \tfrac{1}{2} \times (a + b) \times h$$

5) CIRCLE

DON'T MUDDLE UP THESE TWO CIRCLE FORMULAE!

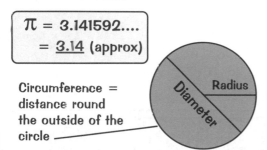

π = 3.141592....
  = 3.14 (approx)

Circumference = distance round the outside of the circle

AREA of *CIRCLE* = π × (radius)²

$$A = \pi \times r^2$$

e.g. if the radius is 3cm, then
A = 3.14×(3×3)
  = 28cm²

CIRCUMFERENCE = π x Diameter

$$C = \pi \times D$$

The Acid Test:

LEARN THIS PAGE — then **COVER THE PAGE AND WRITE DOWN** as much of it as you can **FROM MEMORY**.

Check your effort and *try again!*

Circle Questions

1) π "A Number a Bit Bigger than 3"

The big thing to remember is that π (called "pi") only seems confusing because it's a scary-looking Greek letter. In the end, it's just an _ordinary number_ (3.14159...) _which is rounded off to either_ 3 or 3.14 or 3.142 (depending on how accurate you want to be). And that's all it is: _A NUMBER A BIT BIGGER THAN 3._

2) Diameter is TWICE the Radius

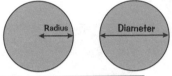

The _DIAMETER_ goes _right across_ the circle.
The _RADIUS_ only goes _halfway_ across.

EXAMPLES:

If the radius is 4cm, the diameter is 8cm,   If D = 16cm, then r = 8cm,
If the radius is 9m, the diameter is 18m,   If diameter = 6mm, then radius = 3mm

3) Arc, Chord and Tangent

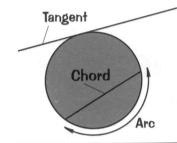

A TANGENT is a straight line that _just touches_ the _outside_ of the circle.

A CHORD is a line drawn _across the inside_ of a circle.

AN ARC is just _part of the circumference_ of the circle.

4) The Big Decision: _"Which circle formula do I use?"_

WORKING OUT <u>AREA</u> OR <u>CIRCUMFERENCE</u> — there is a difference you know!

1) If the question asks for "_the area of the circle_",

   YOU MUST use the FORMULA FOR AREA: $A = \pi \times r^2$

2) If the question asks for "_circumference_" (the distance around the circle)

   YOU MUST use the FORMULA FOR CIRCUMFERENCE: $C = \pi \times D$

AND REMEMBER, it makes _no difference at all_ whether the question gives you _the radius_ or _the diameter_, because it's dead easy to work out one from the other.

EXAMPLE: "Find the circumference and the area of the circle shown below."

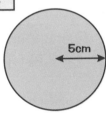

ANSWER: Radius=5cm, so _Diameter=10cm_ (easy)

Formula for CIRCUMFERENCE is:
C = π × D, so
C = 3.14 × 10
 = _31.4cm_

Formula for AREA is:
A = π × r²
 = 3.14 × (5×5)
 = 3.14 × 25 = _78.5cm²_

The Acid Test: _There are 4 SECTIONS on this page._
They're all mighty important — LEARN THEM.

Now cover the page and _write down_ everything you've learnt. Frightening isn't it.
1) A wheel has a diameter of 1m. Find the area and the circumference of it using the methods you've just learnt. Remember to show all your working out.
2) A circular table has a radius of 60cm. Find the area and circumference of it.

Perimeters and Areas

1) Perimeters of Complicated Shapes

Make sure you know these *nitty gritty details* about perimeter:

1) Perimeter is the distance *all the way around the outside of a 2D shape*.

2) To find a **PERIMETER**, you **ADD UP THE LENGTHS OF ALL THE SIDES** , but....
THE ONLY RELIABLE WAY to make sure you get *all the sides* is this:

> 1) <u>Put a big blob at one corner</u> and then go around the shape.
> 2) <u>Write down the length of every side as you go</u>.
> 3) <u>Even sides that seem to have no length given</u> — you must *work them out*.
> 4) Keep going until you get back to the <u>BIG BLOB</u>.

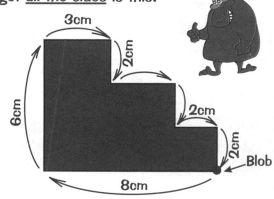

e.g. 8+6+3+2+3+2+2+2 = <u>28cm</u>

Yes, I know you think it's <u>yet another fussy method</u>, but believe me, it's so easy to miss a side. <u>You must use GOOD RELIABLE METHODS for EVERYTHING</u> — or you'll lose marks willy nilly.

2) Areas of Complicated Shapes

> 1) <u>SPLIT THEM UP</u> into *the 3 basic shapes*:
>   RECTANGLE, TRIANGLE, <u>AND CIRCLE</u>.
> 2) <u>Work out the area of each bit SEPARATELY</u> .
> 3) Then <u>ADD THEM ALL TOGETHER</u>
>   (or sometimes SUBTRACT them).

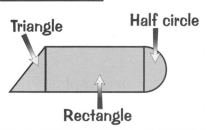

EXAMPLE: *Work out the area of this shape*:

<u>ANSWER</u>:

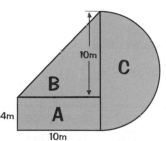

<u>Rectangle A</u>:

Area = l × w
    = 4 × 10
    = <u>40m²</u>

<u>Triangle B</u>:

Area = ½×b×h
    = ½×10×10
    = <u>50m²</u>

<u>Semicircle C</u>:

Area = (π × r²)÷2
    =(3.14 ×7²)÷2
    = <u>76.93m²</u>

<u>TOTAL AREA</u> = 40 + 50 + 76.93 = <u>166.93 m²</u>

The Acid Test: <u>LEARN THE RULES</u> for finding the <u>perimeter and area</u> of <u>complicated shapes</u>.

1) <u>*Turn over and write down*</u> what you've learnt.

2) Find the perimeter and area of the shape shown here:

9cm

5cm

3cm

Volume or Capacity

VOLUMES — YOU MUST LEARN THESE TOO!

1) CUBOID (RECTANGULAR BLOCK)

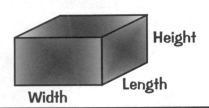

Height
Length
Width

Volume of Cuboid = length × width × height

$$V = l \times w \times h$$

(The other word for volume is *CAPACITY*)

2) PRISM

A PRISM is a solid (3-D) object which has a **constant area of cross-section** — i.e. it's the same shape all the way through.

Now, for some reason, not a lot of people know what a prism is, but they come up all the time in Exams, so make sure YOU know.

Circular Prism
(or Cylinder)

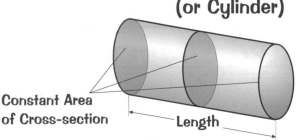

Constant Area
of Cross-section

Length

Hexagonal Prism
(a flat one, certainly, but still a prism)

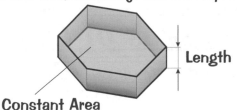

Length

Constant Area
of Cross-section

Triangular Prism

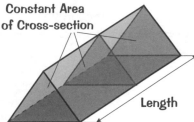

Constant Area
of Cross-section

Length

| Volume of prism | = | Cross-sectional Area | × length |

$$V = A \times l$$

As you can see, the formula for the volume of a prism is *very simple*. The *difficult* part, usually, is *finding the area of the cross-section*.

The Acid Test:

LEARN this page. Then turn over and try to write it all down. Keep trying until you can do it.

Practise these two questions until you can do them all the way through without any hesitation. Name the shapes and find their volumes:

a)
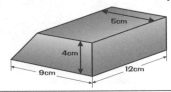
5cm
4cm
9cm
12cm

b)
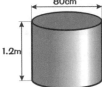
80cm
1.2m

Solids and Nets

You need to know what _Face_, _Edge_ and _Vertex_ mean:

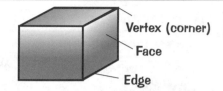

Vertex (corner)
Face
Edge

Surface Area And Nets

1) <u>SURFACE AREA</u> only applies to solid 3D objects, and it's simply _the total area of all the outer surfaces added together_. If you were painting it, it's all the bits you'd paint!

2) There is <u>never a simple formula</u> for surface area — _you have to work out each side in turn and then_ <u>ADD THEM ALL TOGETHER</u>.

3) <u>A NET</u> is just <u>A SOLID SHAPE FOLDED OUT FLAT</u>.

4) So obviously : <u>SURFACE AREA OF SOLID = AREA OF NET</u>.

There are 4 nets that you need to know really well for the Exam, and they're shown below. They may well ask you to draw one of these nets and then work out its area.

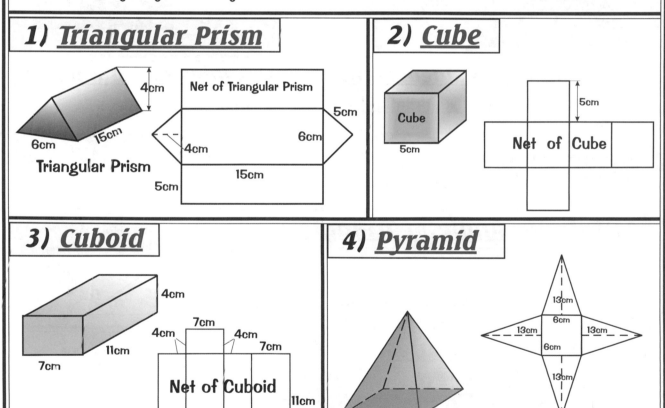

1) _Triangular Prism_

Net of Triangular Prism

4cm
6cm 15cm
Triangular Prism
5cm
4cm
6cm
15cm
5cm

2) _Cube_

Cube
5cm
5cm
Net of Cube

3) _Cuboid_

4cm
4cm 7cm 4cm
7cm 7cm
11cm
Net of Cuboid
11cm

4) _Pyramid_

Square-based Pyramid

13cm
6cm
13cm 13cm
6cm
13cm
Net of Square-based Pyramid

The Acid Test:

LEARN the <u>4 details on Surface Area and Nets</u> and the <u>FOUR NETS</u> on this page, and also the little <u>diagram</u> at the top of the page.

Now cover the page and write down everything you've learnt.
1) Work out the area of all four nets shown above.

Geometry

A couple of definitions to get you started...

PARALLEL LINES
have exactly the
__same slope__   _Parallel lines are
shown by little arrows_

PERPENDICULAR LINES
are at __right angles__ (90°)
to each other
They're shown by a little square:

_This means the angle
is a right angle._

8 Simple Rules — that's all:

If you know them ALL — THOROUGHLY, you at least have a fighting chance of
working out problems with lines and angles. _If you don't — you've no chance._

1) _Angles in a triangle_

Add up to __180°__.

$$a+b+c=180°$$

2) _Angles on a straight line_

Add up to __180°__.

$$a+b+c=180°$$

3) _Angles in a 4-sided shape_

(a "__Quadrilateral__")

Add up to __360°__.

$$a+b+c+d=360°$$

4) _Angles round a point_

Add up to __360°__.

$$a+b+c+d=360°$$

5) _Exterior Angle of Triangle_

Exterior Angle of triangle
= sum of Opposite Interior angles

__i.e.   a+b=d__

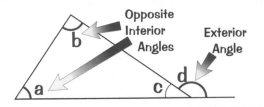

Opposite
Interior
Angles

Exterior
Angle

6) _Isosceles triangles_

__2 sides the same__
__2 angles the same__

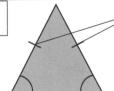

These dashes indicate two
sides the same length

In an isosceles triangle, _YOU ONLY NEED TO KNOW ONE ANGLE_ to be able to find
the other two, which is _very useful IF YOU REMEMBER IT_.

a)

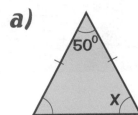

$180° - 50° = 130°$
The two bottom angles are
both the same and they must
add up to $130°$, so
each one must be half of
$130°$ ($= 65°$). So __X = 65°__.

b)

The _two bottom angles must be the_
same, so $55° + 55° = 110°$.
   All the angles add up to $180°$ so
$Y = 180° - 110° = \underline{70°}$.

Geometry

7) Parallel lines

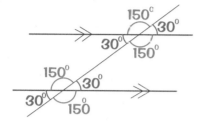

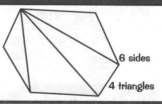

Whenever one line goes across <u>2 parallel lines</u>, then <u>the two bunches of angles are the same</u>

(The arrows mean those 2 lines are parallel — see p68)

> Whenever you have <u>TWO PARALLEL LINES</u> there are *only two different angles:*
> <u>A SMALL ONE</u> and <u>A BIG ONE</u> and they <u>ALWAYS ADD UP TO 180°</u>.
> E.g.   30° and 150°   or   70° and 110°

The trickiest bit about parallel lines is <u>spotting them in the first place</u> — watch out for these "Z", "C", "U" and "F" shapes popping up:

*If necessary, **EXTEND THE LINES** to make the diagram easier to get to grips with:*

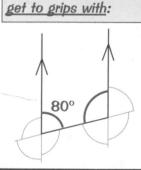

In a <u>Z-shape</u> they're called <u>"ALTERNATE ANGLES"</u>

If they add up to 180 they're called <u>"SUPPLEMENTARY ANGLES"</u>

In an <u>F-shape</u> they're called <u>"CORRESPONDING ANGLES"</u>

Alas you're expected to learn these three silly names too!

8) Irregular Polygons: Interior and Exterior Angles

An irregular polygon is basically any shape with lots of straight sides which aren't all the same. There are two formulas you need to know:

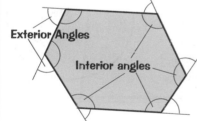

Exterior Angles

Interior angles

> ## Sum of Exterior angles = 360°

> ## Sum of Interior angles = (n − 2)×180°
> where n is the number of sides

The (n − 2)×180° formula comes from splitting the inside of the polygon up into triangles using full diagonals. Each triangle has 180° in it so just count up the triangles and times by 180°. There's always 2 less triangles than there are sides, hence the (n − 2).

6 sides

4 triangles

The Acid Test:
LEARN EVERYTHING on these two pages. Then <u>turn over</u> and see how much of it you can <u>write down</u>.

1) In an isosceles triangle, the angle between the equal sides is 40°.
   What do the other two angles equal?
2) Find the size of angles a and b in the diagram.
3) How much do the interior angles of a 5-sided polygon add up to?
4) One of the diagrams above has one angle given as 80°. Find the other 7 angles.

Circle Geometry

Four Simple Rules — that's all:

You'll have to learn these too if you want to be able to do circle problems.

1) ANGLE IN A SEMICIRCLE = 90⁰

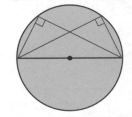

A triangle drawn from the <u>two ends of a diameter</u> will ALWAYS make an <u>angle of 90° where it hits</u> the edge of the circle, no matter where it hits.

2) TANGENT and RADIUS MEET AT 90⁰

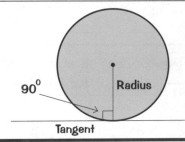

A TANGENT is a line that just touches the edge of a curve. <u>If a tangent and radius meet</u> at the same point, then the angle they make is *EXACTLY 90⁰*.

3) SNEAKY ISOSCELES TRIANGLES
FORMED BY TWO RADII

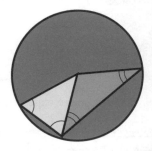

<u>Unlike other isosceles triangles</u> they <u>don't have the little tick marks on the sides</u> to remind you that they're the same — the fact that <u>two sides are radii</u> is enough to make it an isosceles triangle.

4) CHORD BISECTOR IS A DIAMETER

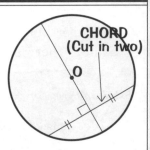

A CHORD is any line <u>drawn across a circle</u>, and no matter where you draw a chord, the line that <u>cuts it exactly in half</u> (at 90⁰), will go <u>through the centre of the circle</u> and so it'll <u>have to be</u> a *DIAMETER*.

5) WHEN YOU'RE STUCK...

It's all too easy to find yourself staring at a geometry problem and <u>getting nowhere</u> — <u>IF SO, this is what you do</u>:

<u>GO THROUGH THE TWELVE RULES, ONE BY ONE</u>, and <u>APPLY EACH OF THEM IN TURN in as many ways as possible</u> — *ONE OF THEM IS BOUND TO WORK.*

The Acid Test:
LEARN these <u>4 Rules</u> plus the 8 Rules on the last two pages. Then <u>turn over and write them all down</u>.

Check your effort and try again — and keep trying till you can do it!

Three-letter Angle Notation

Using Three Letters to Specify Angles

The best way to say which angle you're talking about in a diagram is by using <u>THREE letters</u>. For example in the diagram, <u>angle ACB = 25⁰</u>.

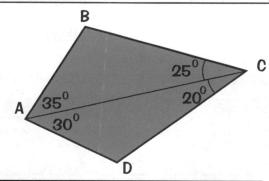

This is the way they'll do it in the Exam so like it or lump it, you'd better get the hang of it. Anyway, it's very simple:

1) The <u>MIDDLE LETTER</u> is <u>where the angle is</u>.

2) The <u>OTHER TWO LETTERS</u> tell you <u>WHICH TWO LINES</u> enclose the angle.

<u>EXAMPLES FROM THE DIAGRAM ABOVE:</u>

1) Angle BCD is <u>AT C</u> and is <u>ENCLOSED BY the lines BC and CD</u> (you just split BCD into BC-CD).   So <u>angle BCD = 45⁰</u>.

2) Angle ACD (AC-CD) is <u>AT C</u> and is <u>ENCLOSED BY the lines AC and CD</u>.   <u>ACD = 20⁰</u>.

A Fairly Tricky Question
— Illustrating the 3-letter notation for Angles

<u>QUESTION</u>:   *"Find all the angles in this diagram".*

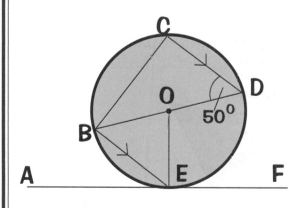

<u>ANSWER:</u>

1) BCD = <u>90⁰</u>   (Angle in a semicircle)

2) So CBO = <u>40⁰</u>   (Angles in triangle = 180⁰)

3) OBE = ODC (=<u>50⁰</u>) (Parallel lines Z shape)

4) OEB = OBE (= <u>50⁰</u>)
    (OEB is a sneaky isosceles)

5) BOE = <u>80⁰</u>
        (Angles in a triangle: 180-50-50=80)

6) OEF = <u>90⁰</u> (Tangent-Radius meet at 90⁰)

7) AEB = <u>40⁰</u> (90⁰-OEB)

The Acid Test:

<u>LEARN</u> what <u>3-Letter notation</u> is. Then <u>cover the page</u> and give an example of it.

1) Looking at the diagram at the top of the page, write down the size of angle BAC and also give the three-letter notation for the angles which are   a) 30⁰ and   b) 65⁰.

2) *Practise the above Example* till you *understand every step* and can do it yourself without any help from the notes — keep practising till you can.

Length, Area and Volume

Identifying Formulas Just by Looking at Them

This isn't as bad as it sounds, since we're only talking about the formulas for 3 things:

LENGTH, AREA and VOLUME

The rules are as simple as this:

> **AREA FORMULAS** always have **LENGTHS MULTIPLIED IN PAIRS**
>
> **VOLUME FORMULAS** always have **LENGTHS MULTIPLIED IN GROUPS OF THREE**
>
> **LENGTH FORMULAS** (such as perimeter) always have **LENGTHS OCCURRING SINGLY**

In formulas of course, <u>LENGTHS ARE REPRESENTED BY LETTERS</u>, so when you look at a formula you're looking for:
<u>GROUPS OF LETTERS MULTIPLIED TOGETHER</u> in *ONES*, *TWOS* or *THREES*.
<u>BUT REMEMBER</u>, π is <u>NOT a length</u>.

Examples:

$6dw + \pi r^2 + 5d^2$ (area)	$6\pi d - 8r/5$ (length)	(r^2 means r x r,
$2d^2w - 5t^3/4$ (volume)	$4\pi r^2 + 9d^2$ (area)	don't forget)
$4\pi r + 12L$ (length)	$Lwh + 8k^3$ (volume)	

<u>Watch out for these last two tricky ones</u>: (Why are they tricky?)

$7c(3c + d)$ (area) $3\pi h(F^2 + 4G^2)$ (volume)

Three Extra Facts:

1) A <u>*QUADRILATERAL*</u> is just *a four-sided shape* — *any* four-sided shape.
So *squares*, *rectangles*, *parallelograms*, etc. are all <u>*QUADRILATERALS*</u>.
And so are the two shown here:

Quadrilaterals

Acute Angle

Obtuse Angle

No, not a cute angel.

2) *ACUTE ANGLES* are *sharp pointy ones* (between $0°$ and $90°$).
3) *OBTUSE ANGLES* are *flatter* (between $90°$ and $180°$).

The Acid Test:

LEARN the <u>Rules for Identifying Formulae</u>, and the <u>Three Extra Facts</u>. Turn over and write it all down.

1) Identify each of these expressions as an area, volume, or perimeter:
πr^2, Lwh, πd, $\frac{1}{2}bh$, $2bh + 3bd$, $4r^2h + \pi d^3$, $2\pi r(2a + 3b)$

Similarity and Enlargement

Congruence and Similarity

Congruence is another ridiculous maths word which sounds really complicated when it's not: If two shapes are *CONGRUENT*, they're simply *the same* — *the same size and the same shape*. That's all it is.

CONGRUENT
— same size, same shape

A, B, and C are CONGRUENT
(with each other)

SIMILAR
— same shape, *different size*

D and E are SIMILAR, (but not congruent)

Remember: when you have *similar* shapes *the angles are always the same*.

Areas and Volumes of Enlargements

Ho ho! This little joker catches everybody out. The increase in area and volume is BIGGER than the scale factor. The rule is this simple:

For an enlargement of Scale Factor n:

The **SIDES** are   n times bigger
The **AREAS** are n^2 times bigger
The **VOLUMES** are n^3 times bigger

Simple... but very forgettable.

FOR EXAMPLE, *if the Scale Factor is 2:*

1) the *lengths* are TWICE AS BIG, (n=2)
2) each *area* is 4 TIMES AS BIG, (n^2=4)
3) the *volume* is 8 TIMES AS BIG, (n^3=8)
   as shown in the diagram:
All YOU have to do is REMEMBER it!

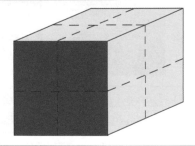

The Acid Test:
LEARN exactly what "**SIMILAR**" and "**CONGRUENT**" mean, and the 3 Rules for INCREASE IN AREA AND VOLUME.

Now cover the page and write down what you've learned.   Then REMEMBER it forever!

1) A tennis ball has a diameter three times bigger than a ping-pong ball. If the volume of the ping-pong ball is 20cm³, what's the volume of the tennis ball?

Enlargements — The 4 Key Features:

1)
If the <u>Scale Factor is BIGGER THAN 1</u> then <u>the shape gets BIGGER</u>.

A to B is an Enlargement, Scale Factor 1½

2)

If the <u>Scale Factor is SMALLER than 1</u> (i.e. a fraction like ½), then the <u>shape gets SMALLER</u>.

(Really this is a *reduction*, but you still call it <u>an Enlargement, Scale Factor ½</u>)

A to B is an Enlargement of Scale Factor ½

3)
Enlargement Scale Factor 3

12cm, 9cm, 4cm, 3cm, 5.5cm, 16.5cm

THE CENTRE OF ENLARGEMENT

The <u>Scale Factor</u> also tells you the <u>RELATIVE DISTANCE</u> of old points and new points <u>from the Centre of Enlargement</u>.

This is <u>VERY USEFUL FOR DRAWING AN ENLARGEMENT</u>, because you can use it to <u>trace out the positions of the new points</u> from the centre of enlargement, as shown in the diagram.

4)
The lengths of the two shapes (big and small) <u>are related to the Scale Factor</u> by this <u>VERY</u> important Formula Triangle <u>WHICH YOU MUST LEARN</u>: (See P.39 on Formula Triangles)

NEW LENGTH / **SCALE FACTOR X OLD LENGTH**

- -

This now lets you to tackle <u>the classic "Enlarged photo" Exam question</u> with breathtaking triviality:

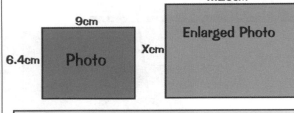

11.25cm, 9cm, 6.4cm, Photo, Enlarged Photo, Xcm

To find the width of the enlarged photo we <u>use the formula triangle TWICE</u>, (firstly to find the <u>Scale Factor</u>, and then to find the <u>missing side</u>):

1) <u>Scale Factor</u> = New length ÷ Old length = 11.25 ÷ 9 = <u>1.25</u>
2) <u>New width</u> = Scale Factor × Old width = 1.25 × 6.4 = <u>8cm</u>

BUT WITHOUT THE FORMULA TRIANGLE YOU'RE SCUPPERED!

The Acid Test:
<u>LEARN</u> the <u>FOUR KEY FEATURES</u> of enlargements, especially the <u>FORMULA TRIANGLE</u>.

Then, <u>when you think you know it</u>, cover the page and <u>write it all down again</u>, from <u>memory</u>, including the sketches and examples, <u>especially the photo enlargement</u> one. Keep trying till you can.

SECTION THREE — SHAPES

The Four Transformations

Translation — ONE Detail **E**nlargement — TWO Details **R**otation — THREE Details **R**eflection — ONE Detail **Y**	1) Use the word _TERRY_ to remember the 4 types. 2) You must always specify _all the details_ for each type.

1) TRANSLATION

You must Specify this ONE detail: 1) The <u>VECTOR OF TRANSLATION</u> $\begin{pmatrix} x \to \\ y \uparrow \end{pmatrix}$

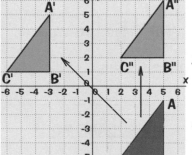

ABC to A'B'C' is a _translation of_ $\begin{pmatrix} -8 \\ 6 \end{pmatrix}$

ABC to A"B"C" is a _translation of_ $\begin{pmatrix} 0 \\ 7 \end{pmatrix}$

2) ENLARGEMENT

You must Specify these 2 details: 1) The <u>SCALE FACTOR</u> 2) The <u>CENTRE</u> of Enlargement

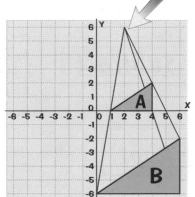

From <u>A to B</u> is an enlargement of <u>scale factor 2</u>, and <u>centre (2,6)</u>

From <u>B to A</u> is an enlargement of <u>scale factor 1/2</u> and <u>centre (2,6)</u>

3) ROTATION

You must Specify these 3 details: 1) <u>ANGLE</u> turned 2) <u>DIRECTION</u> (Clockwise or..) 3) <u>CENTRE</u> of Rotation

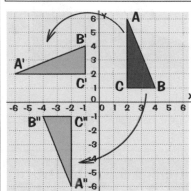

ABC to A'B'C' is a Rotation of <u>90°</u>, <u>anticlockwise</u>, <u>ABOUT the origin</u>.

ABC to A"B"C" is a Rotation of <u>half a turn (180°)</u>, <u>clockwise</u>, <u>ABOUT the origin</u>.

4) REFLECTION

You must Specify this ONE detail: 1) The <u>MIRROR LINE</u>

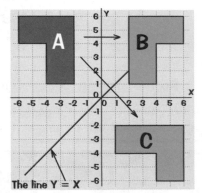

A to B is a <u>reflection IN the Y-axis</u>.

A to C is a <u>reflection IN the line Y=X</u>

The Acid Test:

LEARN the names of the Four Transformations and the details that go with each. When you think you know it, _turn over and write it all down._

1) Describe _fully_ these transformations: A — B, B — C, C — D.

Pythagoras' Theorem

1) <u>PYTHAGORAS' THEOREM</u> goes hand in hand with SIN, COS and TAN because they're both involved with <u>RIGHT-ANGLED TRIANGLES</u>.

2) The big difference is that <u>PYTHAGORAS DOES NOT INVOLVE ANY ANGLES</u> — it just uses *two sides* to find the *third side*. (SIN, COS and TAN always involve <u>ANGLES</u>)

Method The basic formula for Pythagoras' theorem is : $a^2 + b^2 = h^2$

Remember that Pythagoras can only be used on <u>RIGHT-ANGLED TRIANGLES</u>

The trouble is, the formula can be quite difficult to use. *Instead*, it's a lot better to *just remember* these *THREE SIMPLE STEPS*, which work every time:

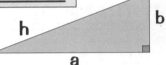

1) Square Them
<u>SQUARE THE TWO NUMBERS</u> that you are given.

2) Add or Subtract
To find the <u>longest side</u>, <u>ADD</u> these two squared numbers.
To find <u>a shorter side</u>, <u>SUBTRACT</u> the smaller one from the larger.

3) Square Root
After adding or subtracting, take the <u>SQUARE ROOT</u>.

Then check that your answer is now <u>SENSIBLE</u>.

Example *"Find the missing side in the triangle shown".*

<u>ANSWER:</u>

Step 1) <u>Square them:</u> $12^2 = 144$, $13^2 = 169$

Step 2) We want to find a <u>shorter side</u>,
         so <u>SUBTRACT</u>: $169 - 144 = 25$

Step 3) <u>Square root:</u> $\sqrt{25} = 5$

So the <u>Missing side</u> = 5m

(You should always ask yourself: "Is it a *sensible answer*?" — in this case you can say "<u>YES</u>, because it's a narrow triangle and about half as long as the other sides")

The Acid Test: LEARN the <u>2 facts</u> relating Pythag. with <u>SIN, COS, TAN</u>, and the <u>3 steps</u> of the Pythag. method.

Now *turn over and write down what you've learned*.

1) If a triangle has one side of 3m and a hypotenuse of 5m, calculate the third side. Then double the length of the sides and check that the theorem still works.

2) Find the missing side PQ.

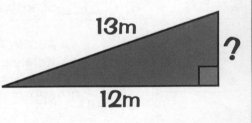

Bearings

Bearings — 3 Key Points

N

The bearing of A from B

●A

B

1) A bearing is the **DIRECTION TRAVELLED** between two points, **GIVEN AS AN ANGLE** in degrees.

2) All bearings are measured **CLOCKWISE** from the **NORTHLINE**.

3) All bearings should be given as **3 figures**, e.g. 045^0 (not 45^0), 316^0, 009^0 (not 9^0), 015^0 etc.

The 3 Key Words

Only learn this if you want to get bearings *RIGHT*

1) "FROM"

Find the word "FROM" in the question, and put your pencil on the diagram at the point you are going "*from*".

2) NORTHLINE

At the point you are going "FROM", *draw in a NORTHLINE*.

3) CLOCKWISE

Now draw in the angle CLOCKWISE *from the northline to the line joining the two points*. This angle is the **BEARING**.

Example

Find the bearing of T from S:

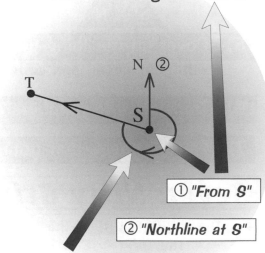

N ②

T

S

① "From S"

② "Northline at S"

③ "*Clockwise*, from the N-line".

This angle is the *bearing of T from S* and is *290^0*.

The Acid Test:

LEARN the 3 Features of Bearings and the 3 Key Steps of the method for finding them.

Now <u>turn over</u> and write down what you've just learnt.
Keep trying <u>till you can write down all six points from memory</u>.

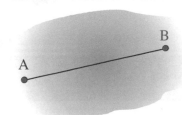

B

A

1) Find the bearing of B from A. (Use a protractor)
2) Find the bearing of A from B.

Trigonometry — SIN, COS, TAN

Using formula triangles to do Trigonometry makes the whole thing *a whole lot easier*, but ALWAYS follow all these steps in this order. If you miss any out you're asking for trouble.

Method

Using SIN, COS and TAN to solve right-angled triangles

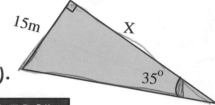

1) Label the three sides O, A and H
   (Opposite, Adjacent and Hypotenuse).

2) Write down FROM MEMORY "SOH CAH TOA"
   (Sounds like a Chinese word, "Sockatoa!")

3) Decide WHICH TWO SIDES are involved O,H   A,H or O,A
   and select SOH, CAH or TOA accordingly

4) Turn the one you choose into a FORMULA TRIANGLE, thus:

(See P.39)

S O H   **C A H**   **T O A**

$\frac{O}{S\theta \times H}$ $\frac{A}{C\theta \times H}$ $\frac{O}{T\theta \times A}$

5) Cover up the thing you want to find
   with your finger, and write down whatever is left showing.

6) Translate into numbers and work it out

7) Finally, check that your answer is sensible.

Seven Nitty Gritty Details

☺ The HYPOTENUSE is the LONGEST SIDE.
   The OPPOSITE is the side OPPOSITE the angle being used (θ).
   The ADJACENT is the side NEXT TO the angle being used (θ).

☺ θ IS A GREEK LETTER called "theta", *and is used to represent ANGLES*

☺ In the formula triangles, Sθ represents SIN θ, Cθ is COS θ ,and Tθ is TAN θ.

☺ To enter, say, SIN 30 into the calculator, just press [SIN] [30] [=] .
   (You should get 0.5 — if you don't, check you're in degrees mode (see P.19)).

☺ Remember, TO FIND THE ANGLE — USE INVERSE (see opposite page →).

☺ ALWAYS USE A DIAGRAM — *draw your own if necessary.*

☺ You can only use SIN, COS and TAN on RIGHT-ANGLED TRIANGLES — you may
   have to *add lines to the diagram to create one* — especially on ISOSCELES triangles.

The Acid Test:
LEARN the 7 Steps of the Method and....
...the 7 Nitty Gritty Details.

Then turn over and write them all down from memory.

Trigonometry — SIN, COS , TAN

Example 1)   *"Find x in the triangle shown."*

1) Label O,A,H
2) Write down "SOH CAH TOA"
3) Two sides *involved*: O,A

4) So use

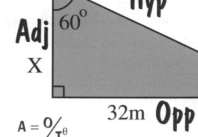

5) We want to find **A** so cover it up to leave: $A = \frac{O}{T\theta}$
6) Translate :

$X = \frac{32}{TAN60}$

Press `32 ÷ TAN 60 =`   `18.475208`   So ans = <u>18.5m</u>

7) Check it's sensible: yes it's about half as big as **32**, as the diagram suggests.

Example 2)   *"Find the angle X in this triangle."*

Note the usual way of dealing with an ISOSCELES TRIANGLE: split it down the middle to get a RIGHT ANGLE:

1) Split the triangle and thus angle **X**. θ=half of **X**.
2) Write down "SOH CAH TOA'"
3) Two sides <u>involved</u>: O,H

4) So use

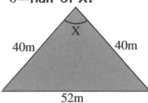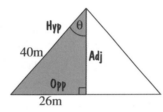

5) We want to find θ so cover up **Sθ** to leave: $S\theta = \frac{O}{H}$

6) Translate:  $SIN\theta = \frac{26}{40} = 0.65$

<u>*NOW USE INVERSE :*</u>   θ = INV SIN (0.65)

Press `INV SIN 0.65 =`   `40.5416`   So θ = <u>40.5°</u>

Since θ= half of **X**,   X=40.5×2   i.e. X=81°.

7) Finally, is it sensible? — Yes, the angle looks like about 80°.

Angles of Elevation And Depression

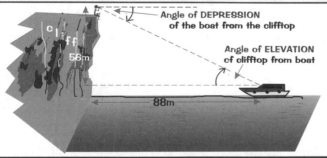

1) The *Angle of Depression* is the angle <u>downwards</u> from the horizontal.

2) The *Angle of Elevation* is the angle <u>upwards</u> from the horizontal.

3) The *Angle of Elevation* and *Angle of Depression* are <u>ALWAYS EQUAL</u>.

The Acid Test:
<u>Practise these questions</u> until you can apply the method <u>fluently</u> and without having to refer to it <u>at all</u>.

1) Find θ 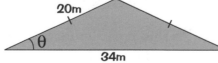   2) Find **X**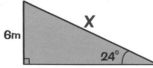

3) Calculate the angles of elevation and depression in the boat drawing above.

SECTION THREE — SHAPES

Loci and Constructions

A <u>LOCUS</u> (another ridiculous maths word) is simply:

> ## A LINE that shows <u>all the points which fit in with a given rule</u>

Make sure you <u>learn</u> how to do these <u>PROPERLY</u> using a <u>RULER AND COMPASSES</u> as shown on these two pages.

1) The locus of points which are <u>"A FIXED DISTANCE from a given POINT"</u>

This locus is simply a <u>CIRCLE</u>.

Pair of Compasses

A given point

The LOCUS of points equidistant from it

2) The locus of points which are <u>"A FIXED DISTANCE from a given LINE"</u>

This locus is an <u>OVAL SHAPE</u>

It has <u>straight sides</u> (drawn with a <u>ruler</u>) and <u>ends</u> which are <u>perfect semicircles</u> (drawn with <u>compasses</u>).

Semicircle ends drawn with compasses

A given line

The LOCUS of points equidistant from it

3) The locus of points which are "EQUIDISTANT from TWO GIVEN LINES"

1) Keep the compass setting <u>THE SAME</u> while you make <u>all four marks</u>.

2) Make sure you <u>leave</u> your compass marks <u>showing</u>.

3) You get <u>two equal angles</u> — i.e. this <u>LOCUS</u> is actually an <u>ANGLE BISECTOR</u>.

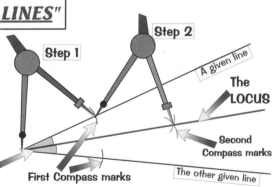

Step 1

Step 2

A given line

The LOCUS

Second Compass marks

First Compass marks

The other given line

4) The locus of points which are "EQUIDISTANT from TWO GIVEN POINTS"

(In the diagram below, A and B are the two given points)

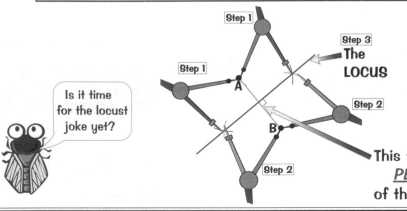

Step 1

Step 1

Step 3
The LOCUS

Step 2

A

B

Step 2

Is it time for the locust joke yet?

<u>This LOCUS</u> is all the points which are the <u>same distance</u> from A and B.

This time the locus is actually the <u>PERPENDICULAR BISECTOR</u> of the line joining the two points.

Loci and Constructions

Constructing accurate 60° angles

1) They may well ask you to draw an _accurate 60° angle_.

2) One place they're needed is for drawing an _equilateral triangle_.

3) Make sure you _follow the method_ shown in this diagram, and that you can do it _entirely from memory_.

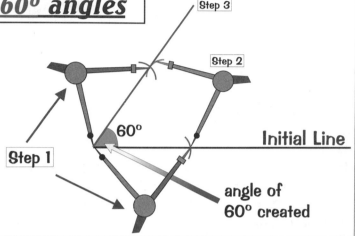

Step 3

Step 2

Step 1

60°

Initial Line

angle of 60° created

Constructing accurate 90° angles

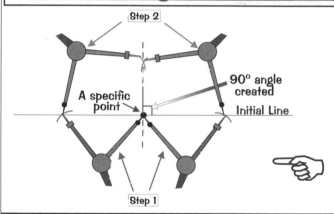

Step 2

A specific point

90° angle created

Initial Line

Step 1

1) They might want you to draw an _accurate 90° angle_.

2) They won't accept it just done "_by eye_" or with a ruler — if you want to get the marks, you've got to do it _the proper way_ with _compasses_ like I've shown you here.

3) Make sure you can _follow the method_ shown in this diagram.

Drawing the Perpendicular from a Point to a Line

1) This is similar to the one above but _not quite the same_ — make sure you can do _both_.

2) Again, they won't accept it just done "_by eye_" or with a ruler — you've got to do it _the proper way_ with _compasses_.

3) _Learn_ the diagram.

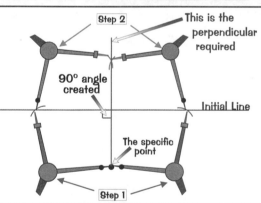

Step 2

This is the perpendicular required

90° angle created

Initial Line

The specific point

Step 1

The Acid Test: LEARN EVERYTHING ON THESE TWO PAGES

Now cover up these two pages and draw an example of each of the four loci. Also draw an equilateral triangle and a square, both with fabulously accurate 60° and 90° angles. Which of the four loci would you need if you had been told:
1) To construct the bisector of an angle;
2) To shade the area nearer to point A than to point B.

Revision Summary for Section Three

I know these questions seem difficult, _but they are the very best revision you can do_. The whole point of revision, remember, is _to find out what you don't know_ and then learn it _until you do_. These searching questions test how much you know _better than anything else ever can_. They follow the sequence of pages in Section Three, so you can easily look up anything you don't know.

Keep learning these basic facts until you know them

1) What is a regular polygon? Draw the first 6, and describe their symmetry.
2) What are the 3 types of symmetry called? Draw an example of each.
3) Draw and name 6 different quadrilaterals and specify all their symmetry.
4) Name 3 different triangles. Draw them and describe their symmetry in full.
5) Write down the formulas for the area of 5 different types of shape.
6) What is π?   What are the two circle formulae?   When do you use them?
7) State 3 important steps for successfully finding the perimeter of a shape.
8) Give the formulas for the volumes of two types of solid.
9) What exactly is a prism? Draw one and show the two important details.
10) Explain what is meant by surface area and what a net is.
11) How are the two related? Is there a formula for working out surface area?
12) Draw the four important nets.
13) List the first 8 rules of geometry, and give extra details about the last 3.
14) Give details of the 4 simple rules for circle geometry.
15) What is the 3-letter notation for angles? Give an example.
16) What are the three rules for identifying formulae as length, area or volume?
17) What is a quadrilateral?
18) Explain what acute and obtuse angles are and give 2 examples of each.
19) Explain what is meant by   a) congruence   b) similarity.
20) If an object is enlarged by a scale factor of 3, how much bigger will the area and volume be for the new shape?
21) What are the rules for this in terms of a scale factor n?
22) In enlargements, what is the effect of a scale factor _bigger_ than 1?
23) What is the effect of a scale factor _smaller_ than 1?
24) How is it used for drawing an enlargement?
25) What is the Formula Triangle for enlargements?
26) Illustrate its use with the enlarged photo question.
27) What does TERRY stand for?
28) Give the details that go with each of the 4 types of transformation.
29) What is the formula for Pythagoras' theorem?
30) Name the three steps of the easy method for doing Pythagoras.
31) What are the three Key Points concerning bearings?
32) Give the three Key Words used to find or plot a bearing.
33) What is trigonometry about? What would a typical question be?
34) How do you decide which sides are the adjacent, opposite and hypotenuse?
35) What is the special word to help you to remember the trigonometry formulae?
36) How do you enter COS 60° into the calculator?
37) What is a locus? Describe in detail the four types you should know.
38) Write down how you would draw accurate 90° and 60° angles.

SECTION THREE — SHAPES

Probability

Probability definitely seems a bit of a "Black Art" to most people. It's not as bad as you think, but <u>YOU MUST LEARN THE BASIC FACTS</u>, which is what we have on these 3 pages.

I'm getting Saturday's draw...

All Probabilities are between 0 and 1

A probability of <u>ZERO</u> means it will <u>NEVER HAPPEN</u>,
A probability of <u>ONE</u> means it <u>DEFINITELY WILL</u>.

You can't have a probability bigger than 1.

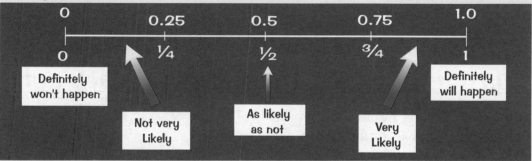

0	0.25	0.5	0.75	1.0
0	¼	½	¾	1

Definitely won't happen

Not very Likely

As likely as not

Very Likely

Definitely will happen

You should be able to put the probability of any event happening on this scale of 0 to 1.

Three Important Details

1) <u>PROBABILITIES SHOULD BE GIVEN</u> as either
   <u>A FRACTION (¼)</u>, or <u>A DECIMAL (0.25)</u>

2) <u>THE NOTATION :</u> "P(x) = ½" <u>SHOULD BE READ AS:</u>
   "<u>The probability of event X happening is ½</u>"

3) <u>PROBABILITIES ALWAYS ADD UP TO 1</u>. This is essential for finding the probability of *the other outcome*. e.g. If P(pass) = ¼, then P(fail) = ¾

THREE IMPORTANT EXTRAS *for combined events (ready for page 86!):*

1) <u>ALWAYS USE YOUR CALCULATOR FRACTION BUTTON</u> for multiplying or adding fractions.

2) Watch out for "<u>WITH REPLACEMENT</u>" and "<u>WITHOUT REPLACEMENT</u>" and make sure you know what difference it makes.
   (*Either you put the thing back after the first go, before having your second go, or you don't — the 2nd tree diagram on page 87 illustrates what can happen*)

3) The <u>COMBINED PROBABILITY</u> of <u>two events</u> <u>BOTH</u> happening is ALWAYS <u>LESS</u> than the probability of either of them occurring alone.

The Acid Test:
<u>LEARN</u> the <u>diagram</u> and the <u>6 IMPORTANT POINTS</u> on this page. Then <u>turn over</u> and <u>write it all down</u>.

1) If P(Picking a red ball) is ⅕, what is the value of P(not picking a red ball)?

Probability in Action

Is it Likely or Not? — know how to Spot It

When you toss a coin, it's as *likely as not* that you'll get a head.
By contrast, if it's a fine summer's day, then it's *not very likely at all* that it'll snow.
You should be able to say *how likely* everyday things are, and *guess at the probability*.

Coloured Balls Have definite Probabilities

If there are 15 blue balls and 5 red balls in a bag, then you can always
calculate exact probabilities.   P (blue ball is selected) $= \frac{15}{20} = \frac{3}{4}$ or 0.75.

If from a different bag, P (green ball is selected) is given as 0.35,
then P (other colour selected) $= 1 - 0.35 = \underline{0.65}$.

With Dice its always a 1/6 chance for each number

When you throw a fair dice twice, the probability of getting 2 sixes is
found by multiplying the two probabilities: $= \frac{1}{6} \times \frac{1}{6}$.

The probability of scoring 11 when 2 dice are thrown is much trickier to work out:

It is the probability of   a 6 *and* then a 5   **OR**   a 5 *and* then a 6

which is $\left(\frac{1}{6} \times \frac{1}{6}\right) + \left(\frac{1}{6} \times \frac{1}{6}\right) = \frac{1}{36} + \frac{1}{36} = \frac{2}{36} = \frac{1}{18}$.

You can draw a *tree diagram* to check this (see P.86 and P.87).

Listing All Possible Outcomes — do it Carefully

In a typical Exam question you might well be asked to list *all the possible outcomes*.
To get them all you've got to do it in a *logical order* — so that you don't miss any.

Example   *"A coin and a dice are tossed together. List all the possible outcomes."*

Start with *one* outcome for the coin and then pair that with *each* outcome for the dice.

e.g.   H 1   H2   H3   H4   H5   H6

Then do the same with the other outcome for
the coin (i.e. tails, T)...

T 1   T2   T3   T4   T5   T6

Tree Diagrams are always a very useful method of finding *all possible outcomes*.

The Acid Test: *LEARN* the four different sections on this page — then *turn over* and *write down everything you've learnt*.

Then try these:
1) If today is a mild October day, how likely is it to be frosty tomorrow?
2) If P (picking a yellow ball) $= \frac{1}{3}$ what is P (not picking a yellow ball)?
3) Find the probability of scoring 3 when two dice are thrown.
4) A black car, a red car and a green car are in a race. List all the possible orders in which the cars can finish.

Estimating Probability

There are _two different ways_ to find out the probability of something happening.

One way is to _calculate it_, like you would for coloured balls in a bag.
This is called _theoretical probability_.

The other way is to do an _experiment_, and simply find out _how many times_ the event _happens_ compared to how many times you _tried_.   This is called _estimated_ probability.

Estimated Probability

Once you've done an experiment and found the estimated probability, you can then estimate the probability of the same event happening in the future, _using the formula below_.

$$P \text{ (Event A)} = \frac{\text{NUMBER OF TIMES EVENT OCCURRED}}{\text{NUMBER OF TRIALS}}$$

EXAMPLE: An _unfair dice_ is thrown 300 times. A "six" is the result 80 times.

Based on this, we can work out: $P(\text{a "six" is thrown}) = \frac{80}{300} = \frac{8}{30} = \frac{4}{15}$.

So $\frac{4}{15}$ is the _estimated_ or _experimental_ probability of getting a six with this _unfair dice_.

You can now use this experimental probability to work out _how many times_ you _expect_ the same thing to happen for a _certain number of tries_. This is the simple formula:

expected number of occurrences = estimated probability × total number of

EXAMPLE: If the unfair dice is thrown _750 times_, how many sixes do you expect to get?

Expected number of sixes = _estimated prob. × no. of tries_ = $\frac{4}{15} \times 750 = \underline{200}$.

Theoretical Probability

Theoretical Probability is the probability that is generally accepted to be a true mathematical chance of something happening.
A _fair die_ has a _theoretical probability_ of $\frac{1}{6}$ for each of the 6 possible results.

So if we throw a _fair_ dice 300 times we would expect to get $\frac{1}{6} \times 300 = \underline{50\ sixes}$.

That's nice and easy.

The Acid Test:
LEARN the 2 formulae to do with estimated probability. Then cover the page and _write down all you have learnt_.

Then try this one:
If a biased coin is tossed 50 times, a head results 35 times. Find the expected number of heads if the coin is tossed 420 times.

Combined Probability

Combined Probability — two or more events

This is where most people start getting into trouble, and d'you know why?
I'll tell you — it's because they don't know these three simple rules:

Three Simple Rules:

1) *Always break down* a complicated-looking probability question into **A SEQUENCE** of **SEPARATE SINGLE EVENTS**.

2) *Find the probability of EACH* of these **SEPARATE SINGLE EVENTS**.

3) *Apply the AND/OR rule:*

1) The AND Rule:

$$P(A \text{ and } B) = P(A) \times P(B)$$

Which means:

The probability of <u>Event A</u> AND <u>Event B</u> BOTH happening is equal to the two separate probabilities **MULTIPLIED** together.

(Strictly speaking, the two events have to be <u>INDEPENDENT</u>. All that means is that one event happening does not in any way stop the other one from happening. Contrast this with *mutually exclusive* below.)

2) The OR Rule:

$$P(A \text{ or } B) = P(A) + P(B)$$

Which means:

The probability of <u>EITHER Event A OR Event B happening</u> is equal to the two separate probabilities **ADDED** together.

(Strictly speaking, the two events have to be <u>MUTUALLY EXCLUSIVE</u> which means that if one event happens, the other one *can't* happen. Pretty much the opposite of *independent events* (see above).)

The way to remember this is that it's the <u>wrong way round</u> — i.e. you'd want the AND to go with the + but it doesn't: It's "<u>AND with ×</u>" and "<u>OR with +</u>".

Example

"Find the probability of picking both red queens from a pack of cards."

<u>ANSWER:</u>

1) <u>SPLIT</u> this into TWO SEPARATE EVENTS
        — i.e. picking one <u>red queen</u> *and then* <u>picking the second red queen</u>.

2) *Find the SEPARATE probabilities* of these *two separate events*:

   P (first red queen) = $\frac{2}{52}$ P (2nd red queen) = $\frac{1}{51}$ (— note the change from 52 to 51)

3) *Apply the AND/OR Rule:*   BOTH events must happen, so it's the AND Rule:

        so <u>multiply</u> the two separate probabilities: $\frac{2}{52} \times \frac{1}{51} = \frac{1}{1326}$.

The Acid Test:

LEARN the <u>Three Simple Rules</u> for <u>multiple events</u>, and <u>the AND/OR Rule</u>.

1) Now turn over and write these rules down <u>from memory</u>. Then apply them to this:
2) Find the probability of picking 2 Kings plus the Queen of Hearts from a pack of cards.

Probability — Tree Diagrams

General Tree Diagram

Tree Diagrams are all pretty much the same, so it's a pretty darned good idea to learn these basic details (which apply to <u>ALL</u> tree diagrams) — ready for the one in the Exam.

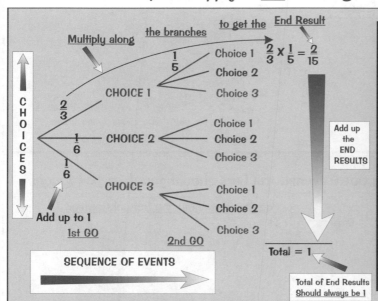

1) Always *MULTIPLY ALONG THE BRANCHES* (as shown) to get the END RESULTS.

2) *On any set of branches which all meet at a point*, the numbers must always ADD UP TO 1.

3) *Check that your diagram is correct* by making sure the End Results ADD UP TO ONE.

4) *To answer any question*, simply ADD UP THE RELEVANT END RESULTS (see below).

A likely Tree Diagram Question

<u>EXAMPLE:</u> "A bag contains 6 red marbles and 4 green marbles. Two marbles are taken <u>without replacement</u>. Draw a tree diagram and hence find the probability that both marbles are the same colour."

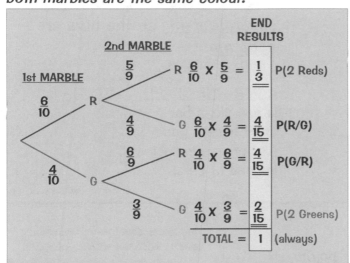

Once the tree diagram is drawn all you then need to do to answer the question is simply <u>select the RELEVANT END RESULTS</u> and then <u>ADD THEM TOGETHER:</u>

2 REDS   (1/3)
2 GREENS   (2/15)

$$\frac{1}{3} + \frac{2}{15} = \frac{7}{15}$$

Use a calculator for this!

The Acid Test:

LEARN the <u>GENERAL DIAGRAM for Tree Diagrams</u> and the <u>4 points</u> that go with them.

1) O.K. let's see what you've learnt shall we:

TURN OVER AND WRITE DOWN EVERYTHING YOU KNOW ABOUT TREE DIAGRAMS.

2) Sam tosses a coin twice. Draw a tree diagram to find the probability that he gets one head and one tail.

Graphs and Charts

Make sure you know all these easy details:

1) Line Graphs or "Frequency Polygons"

A line graph or "frequency polygon" is just a set of points joined up with straight lines.

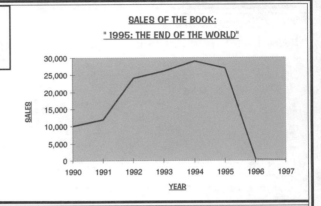

SALES OF THE BOOK:
" 1995: THE END OF THE WORLD"

2) Bar Charts

Just watch out for when the bars should _touch_ or _not touch_:

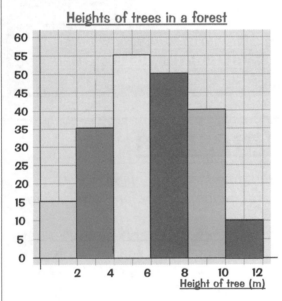

Heights of trees in a forest

Popular Choices at the School Canteen

This bar chart compares _totally separate items_ so the bars are _separate_.

ALL the bars in this chart are for LENGTHS and you must _put every possible length into one bar or the next_ so there mustn't be any spaces.

A BAR-LINE GRAPH is just like a bar chart except you just draw thin lines instead of bars.

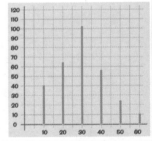

3) Scatter Graphs

1) A SCATTER GRAPH is just a load of points on a graph that _end up in a bit of a mess_ rather than in a nice line or curve.

2) There's a fancy word to say _how much of a mess_ they're in — it's CORRELATION.

3) _Good Correlation_ (or _Strong_ Correlation) means the points _form quite a nice line_, and it means _the two things are closely related to each other_.

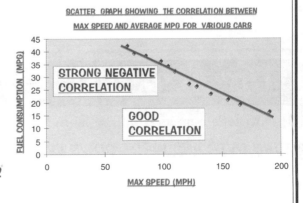

SCATTER GRAPH SHOWING THE CORRELATION BETWEEN MAX SPEED AND AVERAGE MPG FOR VARIOUS CARS

STRONG NEGATIVE CORRELATION

GOOD CORRELATION

Graphs and Charts

Scatter Graphs (continued)

4) _Poor Correlation_ (or _Weak_ Correlation) means the points are _all over the place_ and so there's _very little relation between the two things_.

5) If the points form a line sloping <u>UPHILL</u> from left to right, then there is <u>POSITIVE CORRELATION</u>, which just means that _both things increase or decrease together_.

6) If the points form a line sloping <u>DOWNHILL</u> from left to right, then there is <u>NEGATIVE CORRELATION</u>, which just means that _as one thing increases the other decreases_.

7) So when you're describing a scatter graph you have to mention both things, i.e. whether it's _strong/weak/moderate_ correlation _and_ whether it's _positive/negative_.

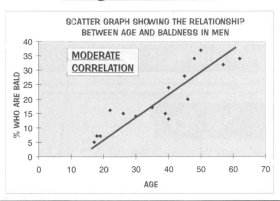

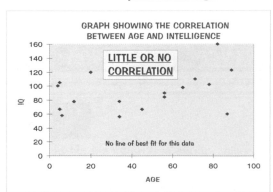

4) Pie Charts

Learn the Golden Rule for Pie Charts:

The TOTAL of Everything = 360°

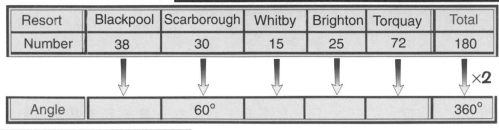

Resort	Blackpool	Scarborough	Whitby	Brighton	Torquay	Total
Number	38	30	15	25	72	180

×2

Angle		60°				360°

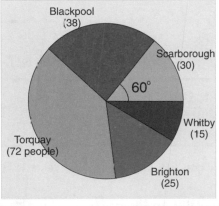

1) Add up all the numbers in each sector to get the <u>TOTAL</u> (← 180 for this one).

2) Then find the <u>MULTIPLIER</u> (or divider) that you need to <u>turn your total into 360°</u>:
For 180 → 360 as above, the <u>MULTIPLIER</u> is 2.

3) Now <u>MULTIPLY EVERY NUMBER BY 2</u> to get the angle for each sector.
E.g. the angle for Scarborough will be
30 × 2 = <u>60°</u> .

The Acid Test:

<u>LEARN THE NAMES</u> of the _four types_ of <u>CHART</u>.

1) Turn over the page and draw an example of each of the 4 charts.
2) Work out the angles for all other holiday resorts in the pie chart shown above.
3) If the points on a scatter graph are all over the place, what does it tell you about the two things that the scatter graph is comparing?

SECTION FOUR — STATISTICS AND PROBABILITY

Mean, Median, Mode and Range

If you don't manage to learn the 4 basic definitions then you'll be passing up on some of the easiest marks in the whole Exam. It can't be *that* difficult can it?

1) MODE   = MOST common

Mode = most (emphasise the 'o' in each when you say them)

2) MEDIAN   = MIDDLE value

Median = mid (emphasise the m*d in each when you say them)

3) MEAN   = TOTAL of items ÷ NUMBER of items

Mean is just the average, "but it's mean 'cos you have to work it out"

4) RANGE   = How far from the smallest to the biggest

I'm mean.

THE GOLDEN RULE:

Mean, median, mode and range should be easy marks but even people who've gone to the incredible extent of learning them, still manage to lose marks in the Exam because they don't do this one vital step:

Always REARRANGE the data in ASCENDING ORDER

(and check you have the same number of entries!)

Example: *"Find the mean, median, mode and range of these numbers:"*

6, -1, 5, -5, 0, 2, 8, 8, -7, 9, 6, 8   (12)

1) **FIRST...** rearrange them: -7, -5, -1, 0, 2, 5, 6, 6, 8, 8, 8, 9   (✓12)

2) MEAN = $\frac{total}{number}$ = $\frac{-7-5-1+0+2+5+6+6+8+8+8+9}{12}$

= 39 ÷ 12 = **3.25**

3) MEDIAN = the middle value (only when they're arranged in order of size, that is!).

When there are TWO MIDDLE NUMBERS, as in this case, then the median is HALFWAY BETWEEN THE TWO MIDDLE NUMBERS

-7, -5, -1, 0, 2, 5, 6, 6, 8, 8, 8, 9
← six numbers this side   ↑   six numbers this side →
Median = 5.5

4) MODE = most common value, which is simply **8**. (Or you can say "The modal value is 8")

5) RANGE = distance from lowest to highest value, i.e. from -7 up to 9, = **16**

The Acid Test: LEARN The Four Definitions and THE GOLDEN RULE...

..then cover this page and write them down from memory.

1) Apply all that you have learnt to find the mean, median, mode and range for this set of data: 11, 21, -10, 0, 2, 18, 16, 4, 11, -21, -5, 25, -7.

Frequency Tables

Frequency Tables can either be done in _rows_ or in _columns_ of numbers and they can be quite confusing, but not if you learn these Eight key points:

Eight Key Points

1) ALL FREQUENCY TABLES ARE THE SAME.

2) The word FREQUENCY just means HOW MANY, so a frequency table is nothing more than a "How many in each group" table.

3) The FIRST ROW (or column) just gives the GROUP LABELS.

4) The SECOND ROW (or column) gives the ACTUAL DATA.

5) You have to WORK OUT A THIRD ROW (or column) yourself.

6) The MEAN is always found using: **3rd Row total ÷ 2nd Row Total.**

7) The MEDIAN is found from the MIDDLE VALUE in the 2nd row.

8) The RANGE is found from the EXTREMES of the first row.

Example

Here is a typical frequency table shown in both ROW FORM and COLUMN FORM:

No. of Letters	Frequency
0	5
1	9
2	8
3	6
4	3
5	0
6	1

No. of Letters	0	1	2	3	4	5	6
Frequency	5	9	8	6	3	0	1

Column Form   _Row form_

There's no real difference between these two forms and you could get either one in your Exam. Whichever you get, make sure you remember these THREE IMPORTANT FACTS:

1) THE 1ST ROW (or column) gives us the GROUP LABELS for the different categories: i.e. "no letters", "one letter", "two letters", etc.

2) THE 2ND ROW (or column) is the ACTUAL DATA and tells us HOW MANY (people) THERE ARE in each category i.e. 5 households had "no letters", 9 households had "one letter", etc.

3) BUT YOU SHOULD SEE THE TABLE AS _UNFINISHED_, because it still needs A THIRD ROW (or column) and TWO TOTALS for the 2nd and 3rd rows, as shown on the next page:

SECTION FOUR — STATISTICS AND PROBABILITY

Frequency Tables

This is what the two types of table look like when they're completed:

No. of letters	0	1	2	3	4	5	6	totals	
Frequency	5	9	8	6	3	0	1	32	(Houses in street)
No. x Frequency	0	9	16	18	12	0	6	61	(Letters delivered)

No. of Letters	Frequency	No. x Frequency
0	5	0
1	9	9
2	8	16
3	6	18
4	3	12
5	0	0
6	1	6
TOTALS	32	61

(Houses in street)   (Letters delivered)

"Where does the third row come from?" ....I hear you cry!

THE THIRD ROW (or column) is ALWAYS obtained by MULTIPLYING the numbers FROM THE FIRST 2 ROWS (or columns).

THIRD ROW = 1ST ROW × 2ND ROW

Once the table is complete, you can easily find the MEAN, MEDIAN, MODE AND RANGE (see P.90) which is what they usually demand in the Exam:

Mean, Median, Mode and Range:

This is easy enough *if you learn it*. If you don't, you'll drown in a sea of numbers.

1) MEAN = $\dfrac{\text{3rd Row Total}}{\text{2nd Row Total}}$ = $\dfrac{61}{32}$ = 1.91 (Letters per household)

2) MEDIAN: — imagine the original data *SET OUT IN ASCENDING ORDER*:

00000 111111111 22222222 333333 444 6
↑

and the median is just the middle which is here between the 16th and 17th digits, So for this data THE MEDIAN IS 2.
(Of course, when you get slick at this you can easily find the position of the middle value straight from the table)

3) The MODE is *very easy* — it's just THE GROUP WITH THE MOST ENTRIES: i.e. 1

4) The RANGE is 6 – 0 = 6 The first row tells us there are houses with anything from "no letters" right up to "six letters" (but not 5 letters). (Always give it as a *single number*)

The Acid Test:

LEARN the 8 RULES for Frequency Tables, then turn over and WRITE THEM DOWN to see what you know.

Using the methods you have just learned and this frequency table, find the MEAN, MEDIAN, MODE and RANGE of the no. of TV's that people have.

No. of TV's	0	1	2	3	4	5	6
Frequency	2	33	28	17	6	3	1

SECTION FOUR — STATISTICS AND PROBABILITY

Grouped Frequency Tables

These are a bit trickier than simple frequency tables, but they can still look deceptively simple, like this one which shows the distribution of weights of a bunch of 60 school kids.

Weight (kg)	31 — 40	41 — 50	51 — 60	61 — 70	71 — 80
Frequency	8	16	18	12	6

Class Boundaries and Mid-Interval Values

These are the two little jokers that make Grouped Frequency tables so tricky.

1) THE CLASS BOUNDARIES are the precise values where you'd pass from one group into the next. For the above table the class boundaries would be at 40.5, 50.5, 60.5, etc. It's not difficult to work out what the class boundaries will be, just so long as you're clued up about it — they're nearly always "something.5" anyway, for obvious reasons.

2) THE MID-INTERVAL VALUES are pretty self-explanatory really and usually end up being "something.5" as well. Mind you a bit of care is needed to make sure you get the exact middle!

"Estimating" The Mean using Mid-Interval Values

Just like with ordinary frequency tables you have to *add extra rows and find totals* to be able to work anything out. Also notice <u>you can only "estimate" the mean from grouped data tables</u> — you can't find it exactly unless you know all the original values.

> 1) <u>Add a 3rd row</u> and enter <u>MID-INTERVAL VALUES</u> for each group.
> 2) <u>Add a 4th row</u> and multiply FREQUENCY × MID-INTERVAL VALUE for each group.

Weight (kg)	31 — 40	41 — 50	51 — 60	61 — 70	71 — 80	TOTALS
Frequency	8	16	18	12	6	60
Mid-Interval Value	35.5	45.5	55.5	65.5	75.5	—
Frequency × Mid-Interval Value	284	728	999	786	453	3250

1) <u>ESTIMATING THE MEAN</u> is then the usual thing of <u>DIVIDING THE TOTALS</u>:

$$\text{Mean} = \frac{\text{Overall Total (Final Row)}}{\text{Frequency Total (2nd Row)}} = \frac{3250}{60} = \underline{54.2}$$

2) THE MODE is still nice'n'easy: <u>the modal group is 51 — 60kg</u> (the one with the most entries).

3) THE MEDIAN can't be found exactly but <u>you can at least say which group it's in</u>. If all the data were put in order, the 30th/31st entries would be in the <u>51 — 60kg</u> group.

The Acid Test:

LEARN all the details on this page, then <u>turn over</u> and <u>write down everything you've learned</u>. Good clean fun.

1) Estimate the mean for this table:
2) Also state the modal group and the approximate value of the median.

Age (yrs)	21 — 30	31 — 40	41 — 50	51 — 60
Frequency	58	61	46	42

Cumulative Frequency Tables

Usually you'll get a half-finished table and they'll ask you to complete it as a cumulative frequency table. This means adding a third row and filling it in (as shown in the example below). Make sure you know these:

FOUR KEY POINTS

1) **CUMULATIVE FREQUENCY** just means **ADDING IT UP AS YOU GO ALONG**. So each entry in the table for cumulative frequency is just "**THE TOTAL SO FAR**".

2) You have to **ADD A THIRD ROW** to the table — this is just the **RUNNING TOTAL** of the 2nd row.

3) If you're plotting a graph, always plot points using the **HIGHEST VALUE** in each group (of row 1) with the value from row 3. (i.e. plot at the *class boundaries*) i.e. for the example below, plot 13 at 160.5, 33 at 170.5, etc.

4) **CUMULATIVE FREQUENCY** is always plotted up the side of a graph, not across.

Example

"Complete the table below for cumulative frequency:"

Height (cm)	141 – 150	151 – 160	161 – 170	171 – 180	181 – 190	191 – 200	201 – 210
Frequency	4	9	20	33	36	15	3

ANSWER: *Add in the third row* where each entry for row 3 (cumulative frequency) is just "THE TOTAL SO FAR" of the numbers for frequency (row 2).

Height (cm)	141 – 150	151 – 160	161 – 170	171 – 180	181 – 190	191 – 200	201 – 210
Frequency	4	9	20	33	36	15	3
Cumulative Frequency	4 (AT 150.5)	13 (AT 160.5)	33 (AT 170.5)	66 (AT 180.5)	102 (AT 190.5)	117 (AT 200.5)	120 (AT 210.5)

The graph is plotted from these pairs: (150.5, 4) (160.5, 13) (170.5, 33) (180.5, 66) etc. because the cumulative frequency has only reached those values (4, 13, 33 etc.) by the **TOP END** of each group, not at the middle of each group, and *150.5* is the actual *CLASS BOUNDARY* between the first group and the next — a tricky detail.

The Acid Test:

LEARN the 4 Key Points, then **turn over** and **write them down**.

1) Complete the table shown here for cumulative frequency.

Height (cm)	151 – 160	161 – 170	171 – 180	181 – 190
Frequency	5	18	21	6

The Cumulative Frequency Curve

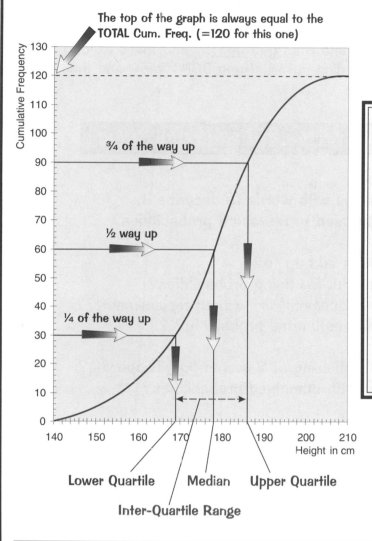

The top of the graph is always equal to the TOTAL Cum. Freq. (=120 for this one)

¾ of the way up

½ way up

¼ of the way up

Lower Quartile   Median   Upper Quartile

Inter-Quartile Range

From the cumulative frequency curve you can get THREE VITAL STATISTICS:

1) MEDIAN
Exactly halfway UP, then across, then down and *read off the bottom scale.*

2) LOWER AND UPPER QUARTILES
Exactly ¼ and ¾ UP the side, then across, then down and *read off the bottom scale.*

3) THE INTER-QUARTILE RANGE
This is the distance *on the bottom scale* between the lower and upper quartiles.

So from the above cumulative frequency curve, we can easily get these results:

MEDIAN = <u>178cm</u>
LOWER QUARTILE = <u>169cm</u>
UPPER QUARTILE = <u>186cm</u>
INTER-QUARTILE RANGE = <u>17cm</u> (186-169)

Interpreting The Shape

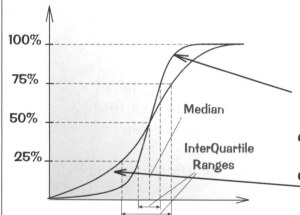

Median

InterQuartile Ranges

The shape of a <u>CUMULATIVE FREQUENCY CURVE</u> also tells us *how spread out* the data values are.

This 'tighter' distribution (which has a small interquartile range) represents very CONSISTENT results, which is usually good — e.g. *lifetimes of batteries or light bulbs* all very close together means a *good product*, compared to the other curve where the lifetimes show *wide variation*, i.e. poor quality product.

The Acid Test:
<u>LEARN THIS PAGE</u>, then <u>cover it up</u> and <u>write down all the important details.</u>

1) Using your completed frequency table from the previous page, draw the cumulative frequency graph and use it to find the three vital statistics.

SECTION FOUR — STATISTICS AND PROBABILITY

Revision Summary for Section Four

Here's the really fun page. The inevitable list of straight-down-the-middle questions to test how much you know. Remember, these questions will sort out quicker than anything else can, exactly what you _know_ and what you _don't_. And that's exactly what revision is all about, don't forget: <u>finding out what you DON'T know</u> and then learning it <u>until you do</u>. Enjoy.

Keep learning these basic facts until you know them

1) How big or small can a probability be?
2) Draw a line to represent all probabilities with words to describe it.
3) Which three types of number can be used to represent probabilities?
4) How should $P(x) = ½$ be read?
5) What must the total probability always add up to?
6) Which calculator button is mighty useful for doing probabilities?
7) What is the full significance of "with replacement" or "without replacement"?
8) What are the two formulae to do with estimated probability?
9) What are combined probabilities?
10) What can you say about the overall probability of 2 events _both_ happening?
11) What are the three rules for dealing with combined probabilities?
12) What is the AND/OR rule?
13) Draw a general tree diagram with all the features that all tree diagrams have.
14) Give the names of the four different types of chart for displaying data.
15) Draw 2 examples of each type of chart.
16) When should the bars of a frequency chart touch and not touch?
17) What does correlation mean? Draw graphs showing the 3 different degrees.
18) What are the 3 steps for finding the angles in a pie chart?
19) Give the definitions for mean, median, mode and range.
20) What is The Golden Rule in connection with mean, median etc.?
21) What are the eight key points for frequency tables?
22) How do you work out the mean and median from a frequency table?
23) How do you find the mode and range from a frequency table?
24) What is the difference between a Frequency Table and a _Grouped_ Frequency Table?
25) What are the 2 things that make Grouped Frequency Tables so tricky?
26) How do you estimate the mean from a grouped frequency table?
27) What are the four key points for cumulative frequency?
28) Do you need to think about _class boundaries_ when plotting a cumulative frequency curve from a table of values? Why?
29) Sketch a typical cumulative frequency graph.
30) What are the 3 vital statistics you can obtain from a cumulative frequency graph?
31) Explain exactly how you obtain them, and illustrate on your graph.
32) How do you decide where halfway up the graph is?

SECTION FOUR — STATISTICS AND PROBABILITY

Answers

SECTION ONE

<u>P1 Multiples, Factors, Prime Factors</u>: 1) 8,16,24,32,40,48,56,64,72,80 and 11,22,33,44,55,66,77,88,99,110
2) 1,2,3,4,6,8,12,24 and 1,2,3,4,5,6,10,12,15,20,30,60 3) 350= 2×5×5×7, 480 = 5×3×2×2×2×2×2.

<u>P2 Prime numbers</u>: 1) 61, 67, 71, 73, 79, 83, 89. 2) 2, 3, 5, 7, 11, 13, 17, 19, 23, 29, 31, 37, 41, 43, 47.

<u>P3 Special Number Sequences</u>:
1) a) EVENS: 2,4,6,8,10,12,14,16,18,20,22,24,26,28,30 b) ODDS: 1,3,5,7,9,11,13,15,17,19,21,23,25,27,29
c) SQUARES: 1,4,9,16,25,36,49,64,81,100,121,144,169,196,225 d) CUBES: 1,8,27,64,125,216,343,512,729,
1000,1331,1728,2197,2744,3375 e) TRIANGLE Nos: 1,3,6,10,15,21,28,36,45,55,66,78,91,105,120
2) 36. 3) a) 1,25,27,49,125 b) 1,16,25,49,64,100,144 c) 1,27,64,125.

<u>P4 Equivalent Fractions</u>: a) 1/2. b) 1/3. c) 1/5. d) 1/8. e) 1/7. f) 1/5. g) 1/4. h) 3/4.

P5 Fractions, Decimals, Percentages:

<u>P6 Rounding Off</u>: 1) 1.07 2) 12.16 3) 90.253 4) 256.0
<u>P7 Rounding Off</u>: 1) a) 7.31 b) 0.06 c) 1.08 d) 4.60
2) a) 0.0358 (Rule 1) b) 63600 (Rule 3) c) 346 (Rule 3) d) 0.710 (Rules 2 & 3).
3) 17 feet 6 in. to 18 feet 6 in.

Fraction	Decimal	Percentage
3/4	0.75	75%
1/5	0.2	20%
7/10	0.7	70%
11/20	0.55	55%
13/20	0.65	65%
7/25	0.28	28%

<u>P9 Accuracy and Estimating</u>: 1) a) Approx 600 miles × 150 miles = 90,000 sq. miles.
b) Approx 7 in cube = 343 cubic inches; 2) a) 2.6kg; b) 183 cm; c) 1000; d) 118mph.
<u>P10 Conversion Factors</u>: 1) 5400m 2) £28 3) 32cm.
<u>P11 Metric and Imperial Units</u>: 1) 88km/h 2) 220 yards 3) 185 cm 4) 70p 5) 11¼ litres.
<u>P13 Fractions</u>: 1 a) 5/28 b) 6/5 c) 2/3 d) 3/5. 2 a) 0.6 b) 22/25 c) 3/10 d) 1⅑ e) 3/8 f) x=13 g) 9.
<u>P15 Percentages</u>: 1) Type 1, £258.50 2) Type 3, £110 3) Type 2, £30, 12.5%.
<u>P19 Calculator Buttons</u>: 1) See P.17 2) [17] [x²] [=] 3) [(-)] [5] [×] [(-)] [8] [=] or [5] [+/-] [×] [8] [+/-] [=]
4) See P.17 5) Fractions 6) [6] [xʸ] [8] [=] 7) [6] [EXP] [8] [=] 8) DEG (or D)
<u>P21 Ratios</u>: 1a) 5:8 b) 2:3 c) 1:2 2) 35 portions 3) £2700 : £600 : £1800.
<u>P23 Standard Index Form</u>: 1) See P.22 2) 9.271 × 10⁵ 3) 2.85 × 10⁻³ 4) 73400 5) 6·6×10¹⁸ 666.....(18 sixes!)·6
<u>P24 Powers</u>: 1) a) 2⁷ b) 3 c) 7⁸ d) 4⁸ e) 6² f) 6⁷ g) 8⁶ ; 2) a) 4 b) 8 c) 11 d) 3 e) 4 f) 5.
<u>P25 Square and Cube Roots</u>: 1) a) 8.06 b) 6.87 c) 12.25 d) 9.61 2 a) y = 5 b) t = 5 c) 2.

SECTION TWO

<u>P29 Basic Algebra</u>: 1) a) -3x + y - 2 b) 3w + 6k - 12k² + 8 2) a) 8a²b – 6ab³ b) 10f² + 7f – 12
c) 1 – 4x + 4x² 3) a) 6q²r(2r² - 4 + 5qr³) b) 3xy (2x²y - x + 4y²z).
<u>P30 Number Patterns</u>: 1) See P.30; 2) a) 1250, 6250 b) 55, 27.5 c) 17, 23 d) 16, 20.
<u>P31 Finding the nth Term</u>: 1 a) 3n + 2 b) 7 – 10n c) ½n(n+1) d) n² – 2n + 6.
<u>P32 Negative Numbers</u>: 1a) +14 (Rule 1) b) -4 (Rule 2) c) X (Rule 1) d) -5 (Rule 1)
2a) 12 b) -216 c) 0 d) -113. <u>P33 Substitution</u>: 2) 20°C.
<u>P34 Solving Equations the Easy Way</u>: 1) x = 11 2) x = 2. <u>P35 Trial and Improvement</u>: 1) x = 1.7 .
<u>P36 The Balance Method for Equations</u>: 1) x = 3. <u>P37 Solving Equations</u>: 1) a) x = 1 b) x= – 1.
<u>P38 Rearranging Formulae</u>: 1) $C = \frac{5}{9}$ (F – 32), $F = \frac{9}{5}C + 32$ 2) a) m = – 7n/6 b) m = np/(p + n).
<u>P40 Density and Speed</u>: 1) See P.40 2) 12.38 g/cm³ 3) 569.36g 4) P.40 5) Time = 25 hrs Dist = 18.9km
<u>P41 Formulas</u>: 1) 0 hours, 32 minutes 38 sec.
<u>P42 X,Y and Z Coordinates</u>: 1) A(5,2) B(4,0) C(6,-2) D(0,-1) E(-2,-5) F(-5,0) G(-2,3) H(0,3)
2) C(0,3,0) D(0,0,0) E(5,0,2) F(5,3,2) G(0,3,2) H(0,0,2)
<u>P43 Easy Graphs You Should Know</u>: 1) a) y = x b) y = –x c) y = 2 d) y = ½x 2)

<u>P44 Finding The Gradient of a Line</u>: 1)
Gradient = -4

<u>P48 Plotting straight line Graphs</u>:
1)

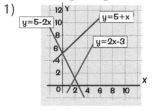

<u>P49</u>: 1)

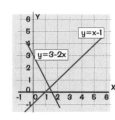

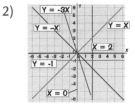

<u>P49</u>: 2) a) y=x+4; b) y=2-3x; c) y=½x+½.

Answers

SECTION TWO (Continued)

<u>P51 Typical Graph Questions</u>: 1)

x	-2	-1	0	1	2	3	4	5	6
y	15	8	3	0	-1	0	3	8	15

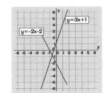

2) Y = 3.8, X =5.6 and -1.6   3) Miles per Gallon, ie. fuel consumption 4) ½ km/h.

<u>P52 Expanding out Brackets</u>: 1) $x^2 + 8x + 7$; 2) $x^2 + 2x - 3$; 3) $x^2 + 4x - 12$; 4) $x^2 - 9x + 20$.

<u>P53 Quadratics</u>: 1) a) X = -4 or 2 b) x = -8 or 3 c) x = 4 or -3 d) x = -7 or 4.

<u>P54 Simultaneous Equations</u>:
V = 6; W = -4 .
<u>P55 Simultaneous Equations</u>:
Solution=(-3/5, -4/5).

<u>P56 Inequalities</u>:
1) X   -4 2) -2, -1, 0 , 1, 2.
<u>P57 Graphical Inequalities</u>:
1)

SECTION THREE

<u>P59 Regular Polygons</u>:1)-4) See P.23 5) Ext. angle = 60^0, Int. angle = 120^0 6) Ext. angle = 36^0, Int. angle = 144^0

<u>P61 Symmetry</u>:

I : 2 lines of symmetry, Rot^{nl}. symmetry Order 2,   N: 0 lines of symmetry, Rot^{nl}. symmetry Order 2
E : 1 line of symmetry, no Rotational symmetry,   Y: 1 line of symmetry, no Rot^{nl}. symmetry
W : 1 line of symmetry, no Rotational symmetry,   Z: 0 lines of symmetry, Rot^{nl}. symmetry Order 2
S : 0 lines of symmetry, Rot^{nl}. symmetry Order 2,   T: 1 line of symmetry, no Rot^{nl}. symmetry

<u>P64 Circle Questions</u>: 1) Area = $0.785m^2$ Circumference = 3.142m   2) A = $11310cm^2$,   C = 377cm

<u>P65 Perimeters and Areas</u>: 2) Perimeter = 36.97cm   Area = $84.31cm^2$.

<u>P66 Volume or Capacity</u>: a) Trapezoidal Prism, V = $336 cm^3$ b) Cylinder, V = $0.60m^3$

<u>P67 Solids and Nets</u>: 1) $264cm^2$ 2) $150cm^2$ 3) $298cm^2$ 4) $192cm^2$.

<u>P69 Geometry</u>: 1) 70° each 2) a=115° b=80° 3) 540° 4) 100° and 80° all round

<u>P71 Three-Letter Angle Notation</u>: 1) BAC = 35° a) DAC=30° b) BAD=65°

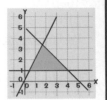

<u>P72 Length, Area and Volume</u>: 1) πr^2 = Area,   Lwh = Volume, πd = Perimeter,
½bh = Area,   2bh + 3bd = Area, $4r^2h + \pi d^3$ = Volume, $3\pi r(2a + 3b)$ = Area

<u>P73 Similarity and Enlargement</u>: 1) $540 cm^3$

<u>P75 Transformations</u>: 1) A→B, Translation of $\binom{5}{-1}$;   B→C, Reflection in the X-axis

C→D, Rotation of 90^0 clockwise about the origin.

<u>P76 Pythagoras</u>: 1) 4m; 6, 8, 10 also right angle triangle *works*. 2) PQ=24m.

<u>P77 Bearings</u>: 1) 75^0 2) 255^0 <u>P79 Trigonometry</u>: 1) 31.79^0 2) X = 14.8m   3) 32.5^0 (both)

<u>P81 Loci and Constructions</u>: 1) Equidistant from two given lines 2) Equidistant from two given points

SECTION FOUR

<u>P83/84 Probability</u>: 1) 4/5   <u>P84</u>: 1) Not very likely 2) 2/3 3) 1/18 4) BRG; BGR; RBG; RGB; GBR; GRB.

<u>P85 Estimating Prob</u>: 294   <u>P86 Combined Prob</u>: 2) 1/11050   <u>P87 Tree Diagrams</u>: 2) 1/2

<u>P89 Graphs and Charts</u>: 2) Blackpool 76^0; Whitby 30^0; Brighton 50^0; Torquay 144^0.

3) That they are not related to each other, i.e. no correlation.

<u>P90 Mean, Median, Mode and Range</u>: First, do this: -21, -10, -7, -5, 0, 2, 4, 11, 11, 16, 18, 21, 25 (13 numbers)

Then you get: Mean = 5, Median = 4,   Mode =11, Range = 46

<u>P92 Frequency Tables</u>:

No. of TV's	0	1	2	3	4	5	6	TOTALS
Frequency	2	33	28	17	6	3	1	90
No. × Frequency	0	33	56	51	24	15	6	185

Mean = 2.1, Median = 2, Mode = 1, Range = 6

<u>P93 Grouped Frequency Tables</u>:

Age (yrs)	21 — 30	31 — 40	41 — 50	51 — 60	TOTALS
Frequency	58	61	46	42	207
Mid-Interval Value	25.5	35.5	45.5	55.5	—
Freq × M I V	1479	2165.5	2093	2331	8068.5

Mean = 38.98, Modal Group = 31 — 40, Median ≈ 31 —40

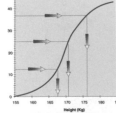

<u>P94/95 Cumulative Frequency</u>:

Height (cm)	151 – 160	161 – 170	171 – 181	181 – 190
Frequency	5	18	21	6
Cum. Freq.	5	23	44	50

Median = 171.3cm,
Lower Quartile = 165.5cm
Upper Quartile = 176cm
Inter-quartile range 10.5cm

Index

Index

MHR39